SpringerBriefs in Earth Sciences

More information about this series at http://www.springer.com/series/8897

Miguel Ortega-Sánchez · Rafael J. Bergillos
Alejandro López-Ruiz · Miguel A. Losada

Morphodynamics of Mediterranean Mixed Sand and Gravel Coasts

 Springer

Miguel Ortega-Sánchez
Andalusian Institute for Earth System
   Research
University of Granada
Granada
Spain

Rafael J. Bergillos
Andalusian Institute for Earth System
   Research
University of Granada
Granada
Spain

Alejandro López-Ruiz
Department of Aerospace Engineering
   and Fluid Mechanics
University of Seville
Seville
Spain

Miguel A. Losada
Andalusian Institute for Earth System
   Research
University of Granada
Granada
Spain

ISSN 2191-5369   ISSN 2191-5377   (electronic)
SpringerBriefs in Earth Sciences
ISBN 978-3-319-52439-9   ISBN 978-3-319-52440-5   (eBook)
DOI 10.1007/978-3-319-52440-5

Library of Congress Control Number: 2016963200

Printed on acid-free paper

This Springer imprint is published by Springer Nature
The registered company is Springer International Publishing AG
The registered company address is: Gewerbestrasse 11, 6330 Cham, Switzerland

Preface

This Springer Briefs volume on the morphodynamics of Mediterranean mixed sand and gravel coasts aims to present the latest achievements in this field for people who are interested to learn more how this type of environment responds to maritime agents. It is intended for engineers and researchers in the areas of Coastal Engineering, Geomorphology and Physical Oceanography who are concerned with the evolution and sustainable management of the coast and for postgraduate students. Coastal environments are some of the most dynamic and complex systems on the earth's surface, and their study involves many different disciplines. Research efforts over the last several decades in the fields of theoretical description, numerical modeling, measuring techniques and data analysis of physical processes have resulted in significant incremental advances in the coastal disciplines. Despite these improvements, such complex environments require much more research.

The book focuses on the main coastal processes driving the morphodynamics of Mediterranean mixed sand and gravel coasts: nearshore waves, littoral drift and the response of both the coastline and the beach profile. It begins with an introduction to the Mediterranean environment, the importance of its coasts and the challenges that should be faced during upcoming decades due to the intensification of coastal retreat (Chap. 1). Then, Chap. 2 presents two study zones in southern Spain that have been intensively studied and provide the major results presented in subsequent chapters. Although the role played by waves arriving at the coast has been intensively studied in the past, significant differences appear when comparing wide sandy shelves with narrow complex ones. Chapter 3 is devoted to illustrating this complexity, which is even more enhanced when the coastline is curvilinear or has rhythmic-type features. Since the pioneering works of the last century, it has been well known that littoral drift drives the major changes in the plan shape of the coastline. Nonetheless, much remains unknown about the dynamics of longshore sediment transport when the settings are far from those commonly found on sandy beaches, as in the case of mixed sand and gravel coasts. Chapter 4 introduces these differences and shows how some of these restrictions could be treated. Moreover, the longshore sediment transport patterns and the plan shape evolution of the two study zones selected are analyzed and linked.

The beach's responses to varying wave, wind and water level conditions are of major importance to proper coast management. Swash processes and the evolution of the beach profile on sandy beaches have been intensively studied over the past several decades; however, similar studies on mixed sand and gravel coasts are still necessary. In Chap. 5, we present relevant results that explain the different behaviors of mixed sand and gravel beaches and open a new vision of how to manage global coastal erosion and sea level rise. Although this book is mainly focused on these results and the corresponding publications, a large list of some of the most relevant and up-to-date references is provided in each chapter. We apologize for any errors that may be found in the book, despite our efforts to eliminate them.

Finally, we would like to highlight that the study of these Mediterranean mixed sand and gravel coasts began with the European Research Project "Human interaction with large scale coastal morphological evolution", led by the last author, Prof. Miguel A. Losada, who also founded the Environmental Fluid Dynamics Group of the University of Granada. His conception of coastal morphodynamics has always been a breakthrough in the state of the art. Within this book, we have tried to enhance some of his main lessons, particularly the importance of understanding basic physical mechanisms as a necessary step to provide new research insights and develop advanced techniques. This is of special interest to those who are beginning work in the discipline, and transmitting it throughout the book was one of the authors aims. The authors are grateful to the members of the Environmental Fluid Dynamics Group of the University of Granada, particularly to Pedro J. Magaña for his assistance in processing the text and figures. We are indebted to Asunción Baquerizo, Francisco J. Lobo and Gerd Masselink for their valuable contributions to the book.

The present work would have not been possible without the funding provided by the European Commission (Contract EVK3-CT-2000-00037) and the Spanish Government through different Research Projects (REN2002-01038/MAR, CTM2005-06853/MAR, CTM2012-32439 and BIA2015-65598-P), Contracts (AP2009-2984 and BES-2013-062617) and Grants (AP2001-4121, EEBB-I-15-10002 and EEBB-I-16-11009). We would also like to thank the support provided by the Andalusian Regional Government (Project P09-TEP-4630) and by the Directorate General for Coasts (Spanish Ministry of Environment).

Granada, Spain   Miguel Ortega-Sánchez
November 2016   Rafael J. Bergillos
                                                                   Alejandro López-Ruiz
                                                                     Miguel A. Losada

Contents

Chapter 1
Introduction

Abstract Within the Mediterranean basin, the genesis of heterogeneous coasts characterized by a mixture of sand and gravel sediment fractions is frequent. The physical processes responsible for their formation and dynamics are different from those of sandy coasts; an important role is played by gravity waves propagating over complex inner shelves. Those inner shelves originated from the tectonic and geomorphic specificities of the Mediterranean basin in combination with a drainage system characterized by rivers with small catchments but with a fundamental impact on coastal evolution. This chapter first introduces some general aspects of the Mediterranean basin and its coasts; then, it presents specific concepts covered in subsequent chapters. Finally, the importance of this study in facing future coastal management challenges is exposed.

1.1 The Mediterranean Basin

The Mediterranean basin is almost 2.6 million km^2 with its overall coastline enclosed by mountainous terrain (Fig. 1.1), except for part of the North African margin [34]. The length is approximately 3,800 km from east to west and 900 km from north to south at its maximum between France and Algeria. The average water depth is approximately 1,500 m, with a maximum depth of 5,121 m (southwestern Greece). It can be divided into two sub-basins: the western and the eastern Mediterranean, which in turn are composed of a series of small basins [34]. The western Mediterranean has an area of approximately 0.9 million km^2 and includes the Alborán Sea (Fig. 1.1a), which allocates the two study zones analyzed in this book (Fig. 1.1b). The Mediterranean Basin comprises a vast set of coastal and marine ecosystems that deliver valuable benefits to all of its coastal inhabitants, including different types of nearshore coastal areas [20].

The evolution of the Mediterranean basin has been analyzed in detailed in the past (i.e., [13, 24]). The present configuration of the Mediterranean zone represents the final stages of a continent-continent collisional orogeny. The resulting crustal context is characterized by active and passive margins, by isostasy and tectonics that have imprinted marked variations in relief, and by present-day tectonics and

© The Author(s) 2017

M. Ortega-Sánchez et al., *Morphodynamics of Mediterranean Mixed Sand and Gravel Coasts*, SpringerBriefs in Earth Sciences, DOI 10.1007/978-3-319-52440-5_1

(a)   **(b)**

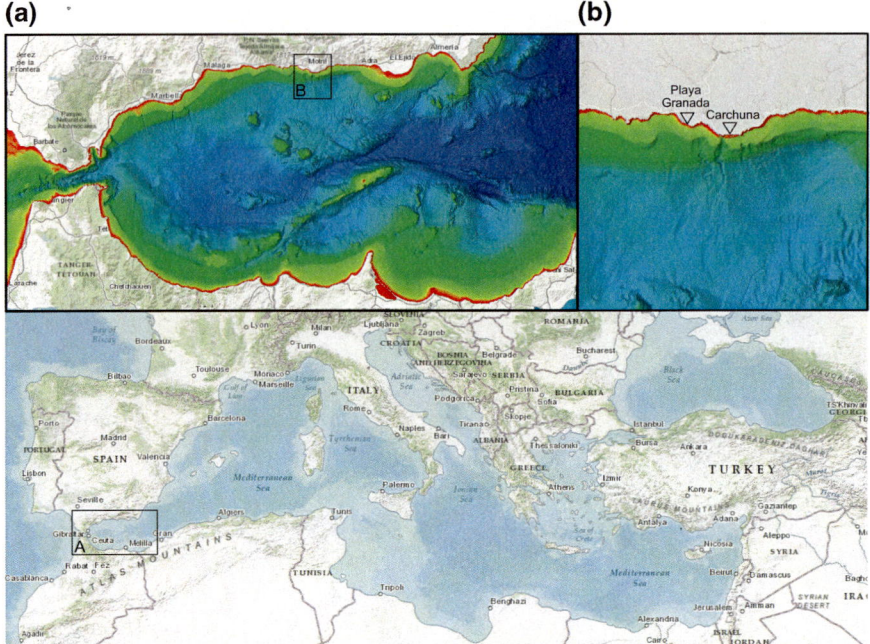

Fig. 1.1 Topographic and bathymetric map of the Mediterranean basin. *Rectangle A* highlights the Alborán basin, whereas *rectangle B* depicts the two study sites analyzed in this book. (*Source* Bathymetry data provided by [14])

coastal geomorphic diversity [2, 11, 13]. The present-day geological configuration of the Mediterranean domain is dominated by a system of connected fold-and-thrust belts and associated foreland and back-arc basins [11]. As a result, small rivers drain approximately 50% of the total Mediterranean Sea catchments [31].

The Mediterranean drainage system comprises more than 160 rivers with individual catchment areas <200 km^2 that bring out the abundance and importance of small rivers in supplying sediment to the coast [2, 7, 20, 21, 23, 39]. The morphology of river channels ranges from confined narrow channels in canyons within upland areas to wide, braided channels in piedmont areas and large valleys [16]. Although numerous Mediterranean river systems formed deltas during the Holocene, during the last several and present centuries, most of the deltas have been experiencing erosion because of human impacts [27]. Consequently, there is a strong dependence between sediment disposal by Mediterranean rivers and the evolution of the coastal zone with a determinant influence on the type of sediments forming the resulting beaches.

Once the sea level stabilized at slightly below its present position, many but not all river basins were able to provide sediment at a rate greater than the removal rate of the combined action of marine agents [27]. The analysis of this relation is complex because of the highly variable river discharges due to climatic disparities and the effect of human transformations [16]. Climate variability also plays a major role in these processes [22].

Along the Mediterranean basin, the continental shelves are generally narrow and mountain slopes drop almost straight into the sea [29]; remarkable examples occur at the southern and northern Iberian Peninsula and at the Maritime Alps (Fig. 1.1). Larger continental shelves are present at locations of seaward extension of deltaic systems, such as off the Ebro and Rhône Rivers [1]. The alongshore supply of riverine sediment has been fundamental for the geomorphic development of these open-coast barrier systems, where coastal morphology and wave fetch conditions favor unimpeded longshore drift [4].

Because of the recent formation of the Mediterranean Sea and the specificities mentioned above, submarine canyons that deeply incise into the continental shelf, others restricted to the slope, others sinuously shaped, and some with roughly linear morphologies are frequently found. Mediterranean canyons can be hundreds of kilometers long, several kilometers wide, and incised into the slope up to 800 m and can have gradients of over 20° on their walls [10]. One of the best examples of continental slope and shelf canyon incision is the Gulf of Lions in the northwestern Mediterranean (Fig. 1.1). As in other areas of the Mediterranean, some of its canyon systems relate directly to Messinian incisions; others are linked to the position of former streams during Plio-Quaternary low stands or originated through retrogressive slides, independent of former features in the shelf; and still others are related to faults or salt tectonics [10]. Moreover, the shape and dimension of these canyons differ substantially from each other. These specific and characteristic elements of the Mediterranean explain the variety of types of beaches found along its coasts (Fig. 1.2).

1.2 Mediterranean Coastline: Beaches

The Mediterranean coastline is approximately 46,000 km long [37] and has nearly 19,000 km of island coastline; 54% of the coastline is rocky, and the remaining 46% includes important and fragile habitats and ecosystems such as beaches, dunes, reefs, lagoons, swamps, estuaries and deltas. Rocky coasts commonly exhibit cliffs cut into different terranes [28] with an increasing occupation by settlements over the last several decades but with a morphology that remains largely unaltered. Conversely, low-lying sedimentary coasts are more dynamic than rocky coasts, and the balance among sea level rise, sediment supply, and wave and coastal current regimes will determine whether the coastline advances, remains stable, or retreats [2, 37, 38].

Although rocky shores and other types of littoral environments are present along the Mediterranean Sea [37], the proportions among them oscillate from one country to another, with Spain exhibiting the lowest bedrock length and Malta the highest. Compared to the Atlantic or Pacific open coasts, wave processes are generally of much lower energy (limited fetch), and the tidal range is generally below a meter (micro-tidal conditions). Therefore, coastline evolution and shoreline features are

Fig. 1.2 Examples of Mediterranean beaches showing the wide range of morphological character-istics. (*Source* Photographs by Miguel Ortega-Sánchez, Pedro J. Magaña and Rafael J. Bergillos)

generally the result of the sediment yield of the rivers, the redistribution of this sediment by nearshore processes and the oscillations of the mean sea level [37].

The Mediterranean comprises not only sand but also gravel barriers associated with the different delta systems that developed in strongly wave-dominated micro-tidal areas [7, 18]. Moreover, embayed shores of different lengths between bedrock headlands are typically found [8, 33]. Those embayed beaches that form in between are generally composed not only of sand but also of a mixture of gravel and cobbles [9, 32]. The dynamics of these environments are highly linked to the sediment supply of nearby rivers because longshore transport is not a primary forcing but a key process in modeling and re-shaping the coast [4]. Therefore, the variety of coastal environments results in a wide range of beaches [6], from sandy to mixed sand-gravel and pure gravel beaches (Fig. 1.2).

1.3 Human Impact

The Mediterranean region has a long history of human settlements and impacts [16, 26]. The total population of the Mediterranean countries grew from 276 million in 1970 to 466 million in 2010 and is predicted to reach 529 million by 2025 [38]. Overall, more than half of the population lives in countries on the southern shores of the Mediterranean, a proportion that is expected to grow to three quarters by 2025 [38]. If it is considered as a single area, the Mediterranean basin is by far the largest global tourism destination, attracting almost a third of the world's international tourists and generating more than a quarter of international tourism receipts.

Anthropogenic modification of sedimentary processes and patterns constitutes both adjustments to natural sedimentary environments (e.g., delta irrigation, coastal reclamation) and the creation of novel sedimentary environments articulated around man-made structures. River damming is needed not only to control extreme events but also to underpin the rise in water demand due to increasing touristic exploitation of the beaches. Many complementary attractions and services were developed to enhance these activities: marinas, promenade waterfronts, resorts and golf campuses, among others. As highlighted by recent works (i.e., [5, 36]), touristic development is placing the littoral zone under unprecedented pressures, resulting in a deep alteration of the natural physical processes that typically control the dynamics of these natural systems.

The direct consequence of this unsustainable pressure on the coast is the development and application of coastal management practices (i.e., [17, 30, 36]), from coastal structures to artificial replenishments. Despite these efforts, the success of many of these practices, mainly artificial nourishments, is temporary and frequently lasts no more than a winter period [6]. Moreover, the sediment that is commonly used to replenish beaches is coarser than the natural sizes, resulting in beaches with different characteristics and morphological responses to forcings.

1.4 Present and Future Challenges

The present and future evolution of the Mediterranean basin in general, and its coasts in particular, depends on several natural factors and human activities [2, 3, 20]. Approximately one-fourth of the European Union Mediterranean coastline suffers from erosion that varies among countries [36], with river damming being the primary reason. The vulnerability of coasts and deltas because of river regulation results in lower sediment supply to the coastal areas and an enhancement of the potential capacity of waves to erode the coast. As already studied in deltas such as Ebro or Guadalfeo (Spain), the reduced supply of sediment to the coast also results in the coarsening of the beachface sediments [6, 30] and in beach morphodynamics changes. The direct consequence is that most of the Mediterranean deltas are no longer advancing but are significantly retreating (i.e., [7, 15, 19, 27, 35]), with a coarser beach sediment composition.

This coastal erosion will increase in the future due to the sea level rise induced by climate change. Climate change model projections for the 21st century predict an increase range up to 61 cm (in a worst-case scenario) for the Eastern Mediterranean [25]. Satellite altimetry data on variations in the level of the Mediterranean Sea between January 1993 and June 2006 indicate that the sea level will rise more in the eastern Mediterranean than in the western Mediterranean. Deltaic areas, because of their topography and sensitive dynamics, are most vulnerable to impacts from sea level rise.

Coastal erosion is a global problem with many environmental impacts on coastal ecosystems. Among the impacts, the integrity of coastal habitats and landscapes and their conservation and planning for further development is probably one of the most important. The approximation to this problem requires a holistic approach such as the one proposed by the Integrated Coastal Zone Management process [12]. The complexity of the problem increases because the drivers and pressures are not uniform along the Mediterranean coasts. An integrated approach is required for the management of activities and conflicts in the coastal zone; however, it is also essential to have a thorough understanding of the key processes that drive and control these coastal environments.

1.5 Objective and Organization of the Book

The main objective of this book is to describe the recent advances achieved in the morphodynamics of mixed sand and gravel beaches on Spanish Mediterranean coasts. This book is organized into 5 chapters, including the Introduction.

Chapter 2 describes the two study zones on the southern coast of Spain (Granada), where many of the advances described in this book where obtained. Carchuna Beach is a 5 km long beach that is characterized by a non-rhythmic cuspate shoreline morphology. The inner shelf exhibits a complex bathymetry with a shelf-indenting submarine canyon, an old river mouth inserted into the bathymetry and different submerged bathymetric undulations. These elements, which occur quite frequently at Mediterranean coasts, control the beach dynamics. The second study zone is Playa Granada, a 3 km long mixed sand and gravel beach located on the side of the Guadalfeo river mouth. Different systematic field measurements were performed at this site to measure both the cross-shore and longshore evolution of the beach profile and the shoreline as well as the specific dynamics of the different sediment fractions. The beach is suffering strong erosion processes after the river regulation that are playing a significant role in the beach dynamics.

Chapter 3 analyzes the impact of the complex inner shelves frequently found along the Mediterranean basin on the nearshore hydrodynamics. By using a calibrated and validated wave propagation numerical model, this chapter first describes the effects of a narrow inner shelf with a shelf-indenting canyon (Carchuna Beach) in the nearshore wave energy distribution. Its nearshore effects result in an alongshore modulation of the nearshore wave energy that controls the shape of the coastline. The presence of submerged undulations in the bathymetry is then analyzed, and they are found to have a significant influence on the morphological processes. Finally, the case of a delta surrounded by a narrow shelf (Playa Granada) is described, highlighting how the delta retreat induced by river regulation modifies the bathymetry and alters the wave propagation patterns.

Chapter 4 is devoted to the impact of the longshore sediment transport on mixed sand and gravel beaches. Traditionally, the coastline is assumed to be quasi-rectilinear, and the formulations are not valid for curved shorelines. This chapter presents a new formulation that not only accounts for the curvature of the shoreline but also includes the variability of the sediment size that is typically found along Mediterranean coasts. It is then applied to the study sites described in the previous chapters to analyze the impact of the hydrodynamics on the morphological response of the coastline.

Finally, Chap. 5 presents the detailed analysis performed by the authors in Playa Granada to characterize the morpho-sedimentary dynamics of this type of beach. The chapter focuses on analyzing the evolution of the beach profile by means of field measurements and data analysis. One of the most significant conclusions is that total run-up (including water-level) plays the major role in dictating the morphological evolution of the beach profile: values of total run-up that are higher than the height of the berm (negative free-board) erode the profile due to overwash processes, whereas positive values of the free-board contribute to its accretion.

References

1. Amblas, D., Canals, M., Lastras, G., Berné, S., Loubrieu, B.: Imaging the seascapes of the Mediterranean. Oceanography **17**, 144–155 (2004)
2. Anthony, E.J., Marriner, N., Morhange, C.: Human influence and the changing geomorphology of Mediterranean deltas and coasts over the last 6000 years: from progradation to destruction phase? Earth Sci. Rev. **139(C)**, 336–361 (2014)
3. Baquerizo, A., Losada, M.A.: Human interaction with large scale coastal morphological evolution. An assessment of the uncertainty. Coast. Eng. **55**(7–8), 569–580 (2008)
4. Bergillos, R.J., López-Ruiz, A., Ortega-Sánchez, M., Masselink, G., Losada, M.A.: Implications of delta retreat on wave propagation and longshore sediment transport—Guadalfeo case study (Southern Spain). Mar. Geol. **382**, 1–16 (2016)
5. Bergillos, R.J., Ortega-Sánchez, M.: Coastal erosion and management practices on a Mediterranean delta, Guadalfeo, southern Spain. Landscape and Urban Planning (under review) (2016)

6. Bergillos, R.J., Ortega-Sánchez, M., Masselink, G., Losada, M.A.: Morpho-sedimentary dynamics of a micro-tidal mixed sand and gravel beach, Playa Granada, Southern Spain. Mar. Geol. **379**, 28–38 (2016)

7. Bergillos, R.J., Rodríguez, C., Millares, A., Ortega-Sánchez, M., Losada, M.A.: Impact of river regulation on a Mediterranean delta: assessment of managed versus unmanaged scenarios. Water Resour. Res. **52**, 5132–5148 (2016). doi:10.1002/2015WR018395

8. Bowman, D., Guillén, J., López, L., Pellegrino, V.: Planview geometry and morphological characteristics of pocket beaches on the Catalan coast (Spain). Geomorphology **108**(3–4), 191–199 (2009)

9. Bramato, S., Ortega-Sánchez, M., Mans, C., Losada, M.A.: Natural recovery of a mixed sand and gravel beach after a sequence of a short duration storm and moderate sea states. J. Coast. Res. **28**(1), 89–101 (2012)

10. Canals, M., Casamor, J.L., Lastras, G., Monaco, A., Acosta, J., Berné, S., Loubrieu, B., Weaver, P., Grehan, A., Dennielou, B.: The role of Canyons in strata formation. Oceanography **17**, 80–91 (2004)

11. Cavazza, W., Wezel, F.C.: The Mediterranean region—a geological primer. Episodes **26**(3), 160–168 (2003)

12. Cicin-Sain, B., Knecht, R.W.: Integrated Coastal and Ocean Management: Concepts and Practices. Island Press, Washington (1998)

13. Comas, M.C., Platt, J.P., Soto, J.I., Watts, A.B.: The origin and tectonic history of the Alboran basin: insights from leg 161 results. In: Zahn, R., Comas, M.C., Klaus, A. (eds.) Ocean Drilling Program, vol. 161, pp. 555–580. Proceedings of the Ocean Drilling Program, Scientific Results (1999)

14. EMODnet Bathymetry Consortium: EMODnet digital bathymetry (DTM). EMODnet Bathymetry (2016). doi:10.12770/c7b53704-999d-4721-b1a3-04ec60c87238

15. Frihy, O.E., Debes, E.A., El Sayed, W.R.: Processes reshaping the Nile delta promontories of Egypt: pre-and post-protection. Geomorphology **54**(3), 263–279 (2003)

16. Hooke, J.M.: Human impacts on fluvial systems in the Mediterranean region. Geomorphology **79**(3–4), 311–335 (2006)

17. Ibáñez, C., Day, J.W., Reyes, E.: The response of deltas to sea-level rise: natural mechanisms and management options to adapt to high-end scenarios. Ecol. Eng. **65**, 122–130 (2014)

18. Jabaloy-Sánchez, A., Lobo, F., Azor, A., Martin-Rosales, W., Pérez-Peña, J.V., Bárcenas, P., Macias, J., Fernández-Salas, L.M., Vázquez-Vilchez, M.: Six thousand years of coastline evolution in the Guadalfeo deltaic system (Southern Iberian Peninsula). Geomorphology **206**, 374–391 (2014)

19. Jimenez, J.A., Sánchez-Arcilla, A.: Medium-term coastal response at the Ebro delta, Spain. Mar. Geol. **114**, 105–118 (1993)

20. Liquete, C., Canals, M., Arnau, P., Urgeles, R., de Madron, X.D.: The impact of humans on strata formation along Mediterranean margins. Oceanography **17**, 70–79 (2004)

21. Lobo, F., Fernandez-Salas, L.M., Moreno, I., Sanz, J.L., Maldonado, A.: The sea-floor morphology of a Mediterranean shelf fed by small rivers, Northern Alboran Sea margin. Cont. Shelf Res. **26**(20), 2607–2628 (2006)

22. Losada, M.A., Baquerizo, A., Ortega-Sánchez, M., Avila, A.: Coastal evolution, sea level and assessment of intrinsic uncertainty. J. Coast. Res. **SI59**, 218–228 (2011)

23. Maetens, W., Vanmaercke, M., Poesen, J., Jankauskas, B., Jankauskiene, G., Ionita, I.: Effects of land use on annual runoff and soil loss in Europe and the Mediterranean: a meta-analysis of plot data. Prog. Phys. Geogr. **36**(5), 599–653 (2012)

24. Maldonado, A., Comas, M.C.: Geology and geophysics of the Alboran Sea: an introduction. Geo-Mar. Lett. **12**(2–3), 61–65 (1992)

25. Marcos, M., Tsimplis, M.N.: Comparison of results of AOGCMs in the Mediterranean Sea during the 21st century. J. Geophys. Res. Oceans **113**, C12028 (2008). doi:10.1029/2008JC004820

26. Marriner, N., Morhange, C.: Geoscience of ancient Mediterranean harbours. Earth Sci. Rev. **80**(3–4), 137–194 (2007)

27. McManus, J.: Deltaic responses to changes in river regimes. Mar. Chem. **79**(3), 155–170 (2002)

28. Ortega-Sánchez, M., Baquerizo, A., Losada, M.A.: On the development of large-scale cuspate features on a semi-reflective beach: Carchuna beach, Southern Spain. Mar. Geol. **198**, 209–223 (2003)
29. Ortega-Sánchez, M., Lobo, F.J., López-Ruiz, A., Losada, M.A., Fernández-Salas, L.M.: The influence of shelf-indenting canyons and infralittoral prograding wedges on coastal morphology: the Carchuna system in Southern Spain. Mar. Geol. **347**, 107–122 (2014)
30. Palanqués, A., Guillén, J.: Coastal changes in the Ebro delta: natural and human factors. J. Coast. Conserv. **4**, 17–26 (1998)
31. Poulos, S.E., Collins, M.B.: Fluviatile sediment fluxes to the Mediterranean Sea: a quantitative approach and the influence of dams. Geol. Soc. Lond. Special Publications **191**, 227–245 (2002)
32. Pranzini, E., Rosas, V., Jackson, N., Nordstrom, K.F.: Beach changes from sediment delivered by streams to pocket beaches during a major flood. Geomorphology **199**, 36–47 (2013)
33. Quevedo, E., Baquerizo, A., Losada, M.A., Ortega-Sánchez, M.: Large-scale coastal features generated by atmospheric pulses and associated edge waves. Mar. Geol. **247**, 226–236 (2008)
34. Rohling, E., Abu-Zied, R., Casford, J., Hayes, A., Hoogakker, B.: The marine environment: present and past. In: Woodward, J.C. (ed.) The Physical Geography of the Mediterranean, pp. 33–68. Oxford University Press (2009)
35. Sabatier, F., Maillet, G., Provansal, M., Fleury, T., Suanez, S., Vella, C.: Sediment budget of the Rhône delta shoreface since the middle of the 19th century. Mar. Geol. **234**, 143–157 (2006)
36. Semeoshenkova, V., Newton, A.: Overview of erosion and beach quality issues in three Southern European countries: Portugal, Spain and Italy. Ocean Coast. Manage. **118**, 12–21 (2015)
37. Stewart, I.S., Morhange, C.: Coastal geomorphology and sea-level change. In: Woodward, J.C. (ed.) The Physical Geography of the Mediterranean, pp. 385–413. Oxford University Press (2009)
38. UNEP/MAP: State of the Mediterranean Marine and Coastal Environment. UNEP/MAP—Barcelona Convention, Athens (2012)
39. Vanmaercke, M., Poesen, J., Verstraeten, G., de Vente, J., Ocakoglu, F.: Sediment yield in Europe: spatial patterns and scale dependency. Geomorphology **130**(3–4), 142–161 (2011)

Chapter 2
Study Sites

Abstract This chapter describes the study sites analyzed in this book: Carchuna and Playa Granada beaches (Southern Spain). Both constitute natural examples of mixed sand and gravel coasts exhibiting the Mediterranean complexity described in the previous chapter. The former site presents complex management practices due to the location of services near the shoreline, and the latter is in a deltaic coast that is strongly affected by river regulation over recent years. Both sites are influenced by a complex narrow inner shelf that enhances the impact of the marine agents on coastal evolution.

The sites are in the southern coast of Spain and face the Mediterranean Sea (Fig. 1.1). The coastline of Carchuna is characterized by non-periodic cuspate features of different dimensions (Fig. 2.1) and a shelf-indenting canyon near the western boundary. Playa Granada can be considered as a regular and quasi-rectilinear beach (Fig. 2.1), whose evolution is strongly dependent on the management of the Guadalfeo River. The sites are widely studied examples of Mediterranean mixed sand and gravel coasts where the processes detailed in this book are highlighted.

2.1 Carchuna Beach

Carchuna Beach is a 4 km long mixed sand and gravel beach located on the Mediterranean coast of Southern Spain and facing the Alborán Sea (Fig. 2.2). The beach is bounded to the west by Sacratif Cape (H1, Fig. 2.2) and to the east by the Punta del Llano Promontory (H6, Fig. 2.2). The beach exhibits a series of large-scale cuspate features with an alongshore spacing of hundreds of meters. These features are bounded by a series of seaward-extending horns (H2–H6, Fig. 2.2).

The short streams which discharge in Carchuna Beach have their sources in the high mountainous relief of the Alpujarrian complex. The hydrological basins of these rivers are characterized by steep relief; the foothills over Carchuna Beach and neighboring areas exceed a 40% slope with non-cohesive material covering the rock substrate. The hillsides show frequent outcrops of rocks as a consequence of the high erosion rate. These streams have concave longitudinal profiles with slopes decreasing toward the mouth [24].

© The Author(s) 2017
M. Ortega-Sánchez et al., *Morphodynamics of Mediterranean Mixed Sand and Gravel Coasts*, SpringerBriefs in Earth Sciences,
DOI 10.1007/978-3-319-52440-5_2

Fig. 2.1 Photographs of Carchuna (*left*) and Playa Granada (*right*) beaches. (*Source* Photographs by Miguel Ortega-Sánchez and Rafael J. Bergillos)

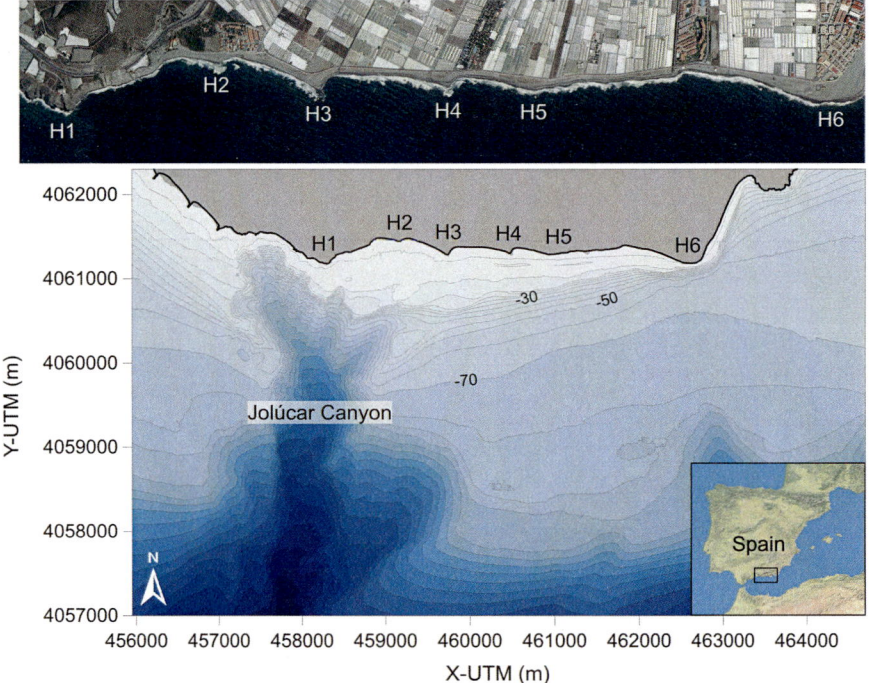

Fig. 2.2 Plan view of Carchuna Beach showing the location of the horns (H1–H6) and detailed bathymetric contours (in meters below the present sea level)

Their main courses extend below the mean sea level approximately down to 75 m depth (Fig. 2.3). These profiles show three different regions: (1) the upper region extending from 300 m altitude down to 40 m; (2) the middle region, where the stream slows down and discharges water and solids into the sea, extending to between 40 and 375 m; and (3) the offshore region where the slope is steep again [24]. As a result of

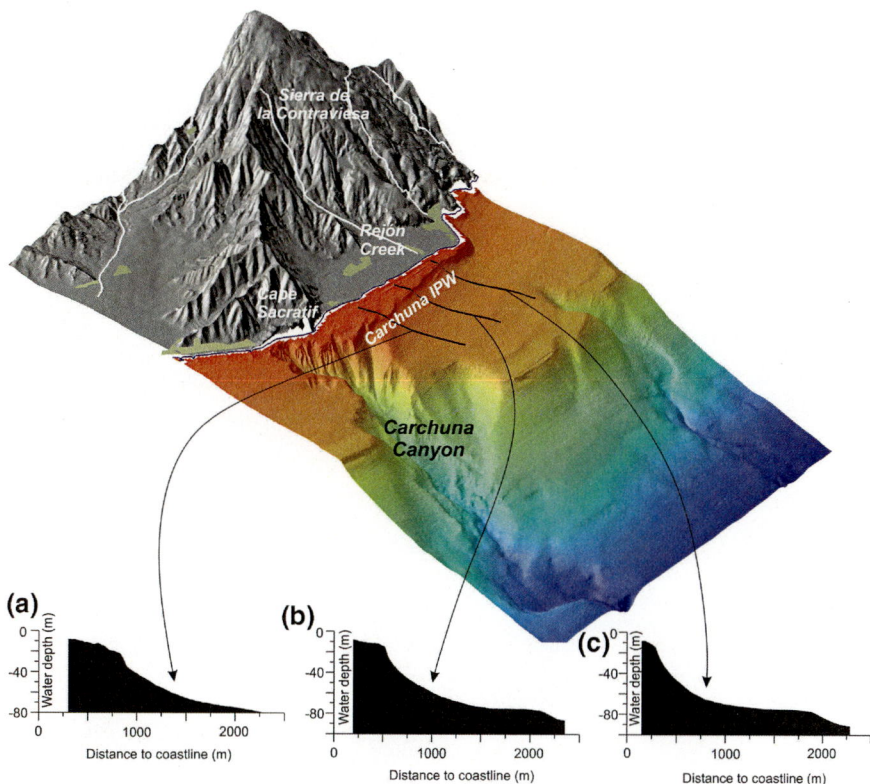

Fig. 2.3 Three dimensional representation of the topo-bathymetric morphology of the Carchuna system. The lateral continuity of the Carchuna IPW is interrupted by the Carchuna Canyon. Its seaward extent decreases from west (up to 1.6 km in the vicinity of the Carchuna Canyon) to east (several hundreds of meters around the Punta del Llano Promontory). (*Source* [26]. Reproduced with permission of Elsevier)

the fluvial and marine processes, the offshore bathymetry opposite Carchuna Beach is characterized by straight and convergent contours toward Carchuna horn, furrowed transversely by several submerged valleys (Fig. 2.3). Chief among these submerged valleys is Carchuna Canyon [26].

Unlike the majority of the submarine canyons of the northern margin of the Alborán Sea [1], the head of Carchuna Canyon is located a short distance from the coastline. The Canyon trends N-S, paralleling the Motril Canyon located to the west [23]. It ends at a water depth of approximately 700 m and has a total length of approximately 5 km. Although the major part of the canyon is located in deep water, the upper part of the western and eastern tributaries cut the shelf (Fig. 2.4). These tributaries, which extend to water depths from approximately 10 m to approximately 180 m, are relatively small in comparison with the main channel of the canyon [26].

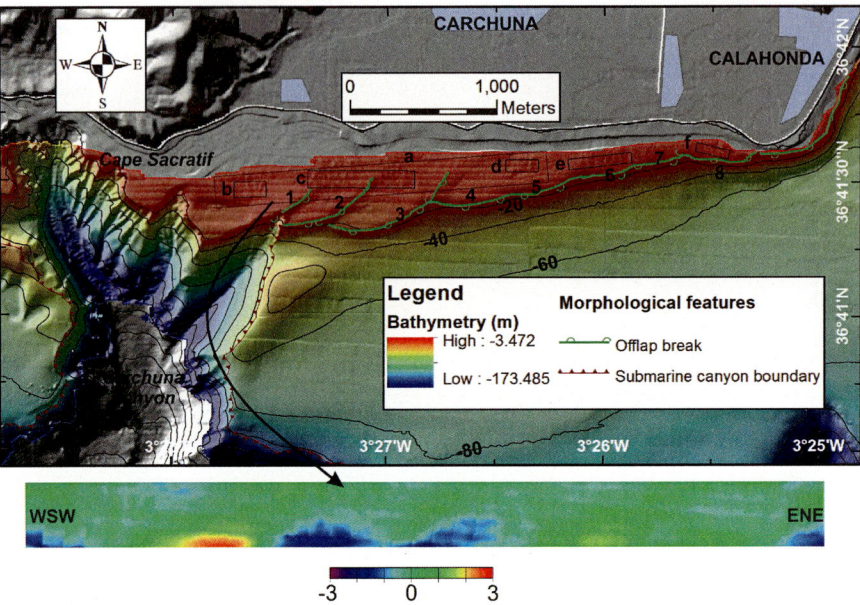

Fig. 2.4 Geomorphological features of the Carchuna IPW, including a marked offlap break and the distal termination. The offlap break (or clinoformrollover) indicates the topset to foreset transition and is composed of a succession of individual breaks of slope, curved in the western part of the system and more rectilinear in the eastern part. The distal boundary is set at the 60 m bathymetric line, where a significant slope reduction is evidenced at the base of the foresets. (*Source* Adapted from [26]. Reproduced with permission of Elsevier)

To the east of Carchuna Canyon, a wedge-shaped sedimentary body exists that has been interpreted as an infralittoral prograding wedge (IPW) in previous studies (Figs. 2.3 and 2.4). The physical continuity between the emerged beach ridge units and several marked increases in slope occurring over the Carchuna IPW has been interpreted as the result of a linked genetic mechanism with oblique or lateral progradation of successive IPWs rather than growth normal to the nearby coastline [10]. The shelf is covered by mainly gravelly sands in the vicinity of the study area. The sand content decreases steadily seaward, ranging from more than 90% at a water depth of less than 20 m (landward of the well-marked increase in slope characterizing the Carchuna IPW) to approximately 70% in the vicinity of the shelf break. In contrast, the gravel content increases steadily from less than 5% at shallow depths to as much as 25% at the distal shelf margin [8].

Analysis of aerial photographs since 1956 and images obtained from a video camera station installed at Sacratif Cape in November 2002 has not revealed significant changes in the alongshore locations of the horns [19]. The beach slope varies along the length of the embayments from 0.04° (in the middle) to 0.3° (in the horns) due to the alongshore variation in the sediment grain size. The steeper-sloping zones are related to larger proportions of coarser sediments [7]. The presence of Carchuna

Canyon and the coarse sediments at the horns suggest that the sediment exchange through the west and east boundaries of the unit is almost negligible. Moreover, the actual sediment supply from the local streams is also negligible, because of the construction of a large number of greenhouses and terraces in their beds to divert water for irrigation. Thus, it can be assumed that there is no sediment input to Carchuna Beach. Because of the large grain sizes, which armor the foreshore, as well as the mild annual wave energy flux, the beach is eroding slowly, but consistently.

2.2 Playa Granada

Playa Granada is a 3 km long mixed sand and beach located on the southern coast of Spain that faces the Mediterranean Sea (Fig. 2.5). The beach corresponds to the

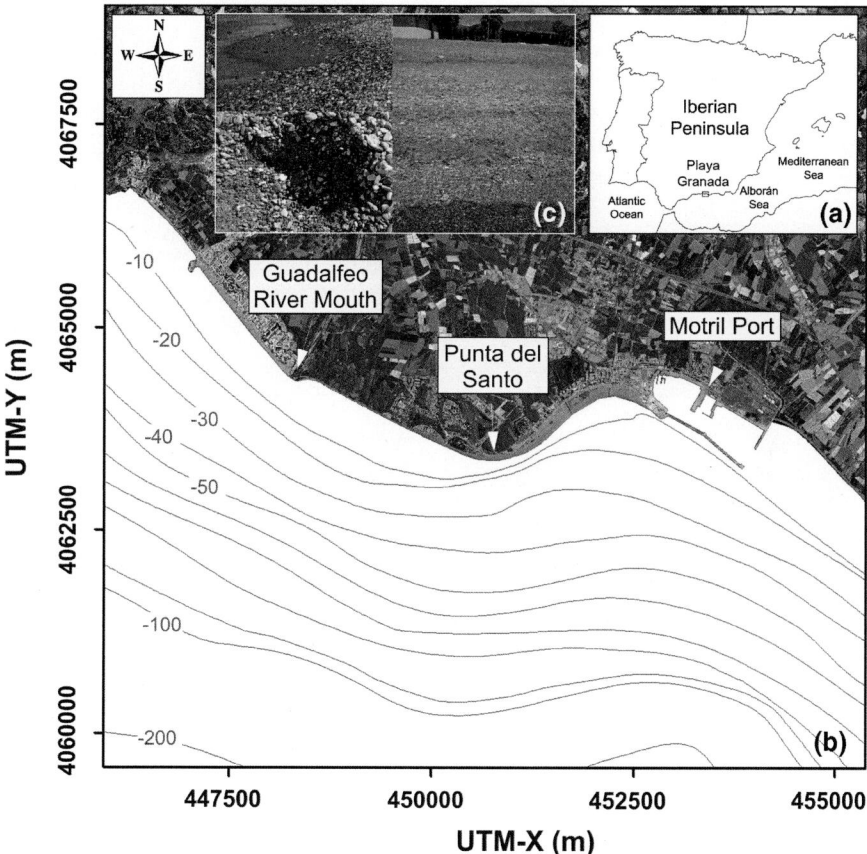

Fig. 2.5 **a** Location of Playa Granada (Southern Spain). **b** Plan view of Playa Granada and bathymetric contours (in meters below the present sea level). **c** Sediment variability on the beach. (*Source* Adapted from [5]. Reproduced with permission of Elsevier)

central stretch of the Guadalfeo deltaic plain [2] and is bounded to the west by the Guadalfeo River mouth and to the east by *Punta del Santo*, the former location of the river mouth. The deltaic coast is bounded to the west by Salobreña Rock and to the east by Motril Port. This port is an artificial barrier that prevents eastward longshore sediment transport [12].

The Andalusian littoral of the Mediterranean Sea is characterized by the presence of high mountainous relief and short fluvial streams, and the main contributor of sediments to the beach is the Guadalfeo River. Its basin has an area of $1,252 \times 10^6$ m^2 and includes the highest peaks on the Iberian Peninsula ($\sim$3,400 m.a.s.l.). Consequently, the river is associated with one of the most high-energy drainage systems along the Spanish Mediterranean coast. The northern catchment divide corresponds to the crest line of the Sierra Nevada, whereas the southern divide corresponds to the crest lines of the Sierra de la Contraviesa and the Sierra de Lújar (Fig. 2.6).

The mountainous influence of the Sierra Nevada conditions the hydrological dynamics and the pluvio-nival character of this semi-arid and high-mountain basin. This sub-basin is mainly composed of Nevado-Filábride complex (mica-schist and graphitic mica-schist). The high altitude guarantees the presence of snow from November to June, which allows a near-perennial flow despite its aridity [13]. The nival contributions condition this quasi-perennial flow, which allows the development of an armor layer and separates a surface layer ($D_{50} \sim 60$ mm) from a substrate layer

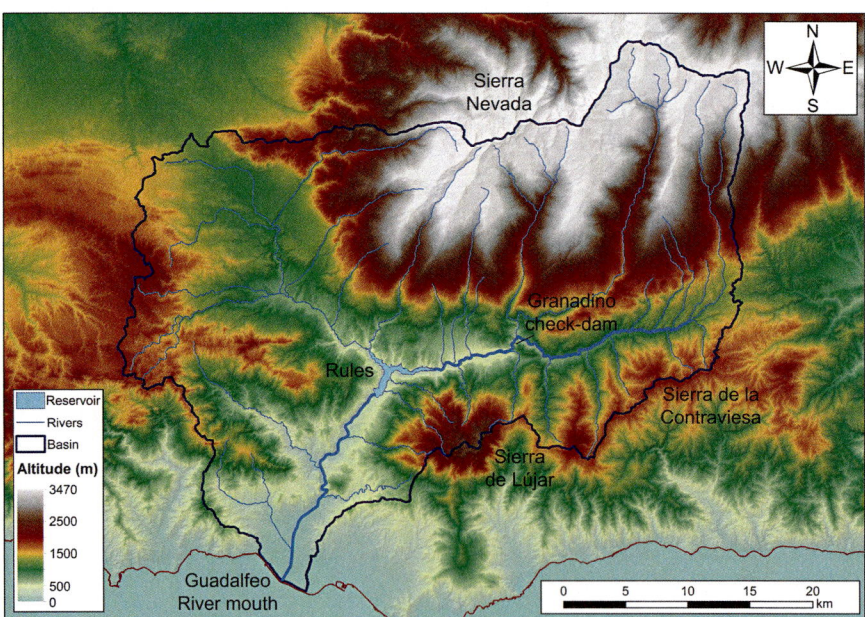

Fig. 2.6 Delimitation of the basin and locations of the Rules' Reservoir, the Granadino check-dam, the Sierra Nevada, the Sierra de la Contraviesa and the Sierra de Lújar

Table 2.1 Minimum, maximum, mean and standard deviation (SD) of the annual precipitation (in mm/y) in 11 meteorological stations of the Guadalfeo basin [21]

Station	Minimum	Maximum	Mean	SD
Albuñol	210	1026	456	179
Bérchules	301	1617	658	249
Cádiar	218	1217	429	186
Contraviesa	312	1062	627	285
Órgiva	207	1434	482	195
Poqueira	434	1485	829	379
Pórtugos	326	1719	726	286
Soportújar	336	1810	692	261
Tajos Breca	413	1474	793	381
Torvizcón	262	1265	539	188
Trevélez	307	1635	663	258

($D_{50} \sim 2.5$ mm). The periodic occurrence of intense precipitation and snowmelt events reshape this drainage network and release a large amount of sediment [22].

The Sierra de la Contraviesa presents a more ephemeral nature with the absence of snowmelt cycles and sub-surface storage. It is composed of Alpujárride complex (quartzites, phyllites and schists). The evolution of erosion processes is clearly influenced by changes in vegetation and land use. This area was originally dominated by forest and Mediterranean shrubs, and large areas of almond and olive orchards are currently pre-eminent [20]. This change has led to the emergence and development of different types of incisions in the form of rills, gullies and more developed channels. Here, the tributary channels lead to important bed-load contributions during intense events that accumulate in the Guadalfeo River.

The annual precipitation data show significant spatial gradients (Table 2.1) and the average annual rainfall in the basin is 586 mm, with minimum and maximum values of 500 and 1000 mm, respectively [20, 21]. The pre-regulation hydrological regime of the Guadalfeo River had peak discharges that exceeded 1,000 m^3/s [9]. The river longitudinal profile is variable: the slope is greater than 2% in the Southern Sierra Nevada and Sierra de la Contraviesa, approximately 1% upstream of the Granadino check-dam, equal to 2.5% between the Órgiva's gauge station and the Rules' Reservoir, and approximately 0.9% downstream of the dam (Fig. 2.6). This relatively steep topographic gradients lead to large contributions from a wide range of sediment sizes [21]. As a result, the particle size distribution on the coast is particularly complex with varying proportions of sand and gravel (Fig. 2.5b). Three sediment fractions are predominant in the studied coastal area: sand ($\sim$0.35 mm), fine gravel ($\sim$5 mm) and coarse gravel ($\sim$20 mm) [4].

The river was dammed 19 km upstream from the mouth in 2004, regulating 85% of the basin runoff [18]. The total capacity of the Rules' Reservoir (117 hm^3) was planned to be used for the following purposes: irrigation (40%), supplies for residential developments along the coast (19%), energy generation (9%), flood control (30%) and environmental flow (2%). However, the river damming modified the natural flow regime and altered the behavior of the system downstream. In particular, the reduction in sediment supply to the coast due to river regulation has been greater than 74,000 m^3/y since 2004 compared to the volume that would have reached the coast under natural conditions [6], contrasting with the accumulation of sediment as delta deposits in the reservoir upstream [22]. As a consequence, the deltaic coast, whose dynamics has been historically governed by the sediment supply of the river during intense events [14, 15], currently presents coastline retreat and severe erosion problems [3].

The studied stretch of beach, which is occupied by an exclusive leisure resort, golf courses, restaurants and summer homes, has been particularly affected and has presented higher levels of coastline retreat in recent years than both western and eastern stretches, known as Salobreña and Poniente Beach, respectively (Fig. 2.7). In addition, Playa Granada has urban lots, at the south of the river mouth, that have not been developed yet. In light of these facts, it is clear that the coast has a high environmental and tourist value, and its exploitation requires a large area of dry beach

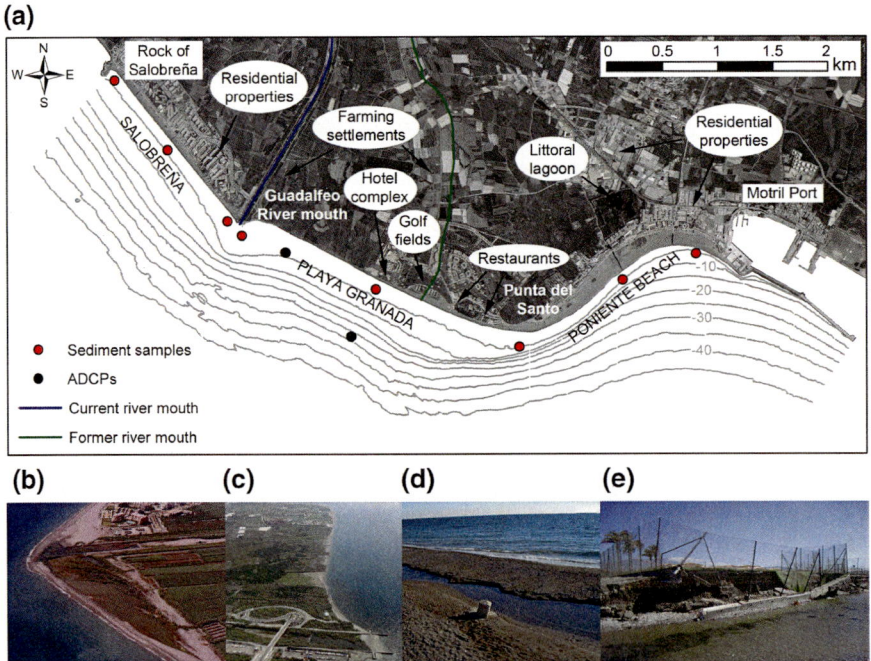

Fig. 2.7 **a** Locations of former and current river mouths, beach profiles where the sediment samples were taken (*red circles*), ADCPs (*black circles*), principal occupations and bathymetric contours in 1999. Plan views of the delta before (**b**) and after (**c**) the river damming. **d** Boundary marker of the public domain that is located few meters from the shoreline due to coastline retreat in Playa Granada. **e** Storm-induced erosion problems in the hotel complex indicated in *panel a*

Table 2.2 Artificial replenishment projects carried out on the coast since the entry into operation of the dam

Year	2006	2009	2010	2014	2014
Volume (m^3)	70,950	51,375	1,654	19,436	106,676
Purpose	Protection	Protection	Tourism	Tourism	Protection

[12]. For this reason, artificial nourishment projects in this mixed sand-gravel coastal environment have been frequent since the entry into operation of the dam (Table 2.2).

The continental shelf of the Guadalfeo River is narrow with an average width of less than 5 km. The shelf break is located at a depth of 100 m and is approximately parallel to the main coastline orientation of the delta front [17]. The shelf gradient is >3° in the delta foreset region and then decreases seaward to <1.5° in the bottomset region [15]. The Guadalfeo River pro-delta extends seaward almost 3.5 km and is characterised by an undulating pro-delta surface due to the presence of bedforms [17]. An off-lap break is identified proximally over the pro-delta, at water depths of 8–14 m and up to 0.5 km from the coast [11]. Medium sands with some muddy intercalations are found in the Guadalfeo pro-delta, whereas the sediment composition in shallower water across the foreset region is dominated by sandy sediments [16]. The emerged deltaic area of the fluvial system covers 8.6×10^6 m^2 and is composed of coarse-grained sediments ranging from medium sands to boulders [15].

2.3   Maritime Climate

Climatic patterns at the study sites exhibit a significant contrast between summer and winter. The region is subjected to the passage of extra-tropical Atlantic cyclones and Mediterranean storms with average wind speeds of 18–22 m/s [24, 27] which generate wind waves under fetch-limited conditions (approximately 300 km). The storm wave climate is distinctly bimodal with the prevailing west-southwest (extra-tropical cyclones) and east-southeast (Mediterranean storms) wave directions (Fig. 2.8). Peak significant wave heights during typical and extreme storm events exceed 2.1 and 3.1 m, respectively. Under South Atlantic storm conditions, swell waves generated in the Gulf of Cadiz propagate through the Strait of Gibraltar. These swell waves impinge the coast simultaneously with the local wind waves, but with slightly different angles [25]. The astronomical tidal range is ~0.6 m (microtidal conditions), whereas typical storm surge levels can exceed 0.5 m [5].

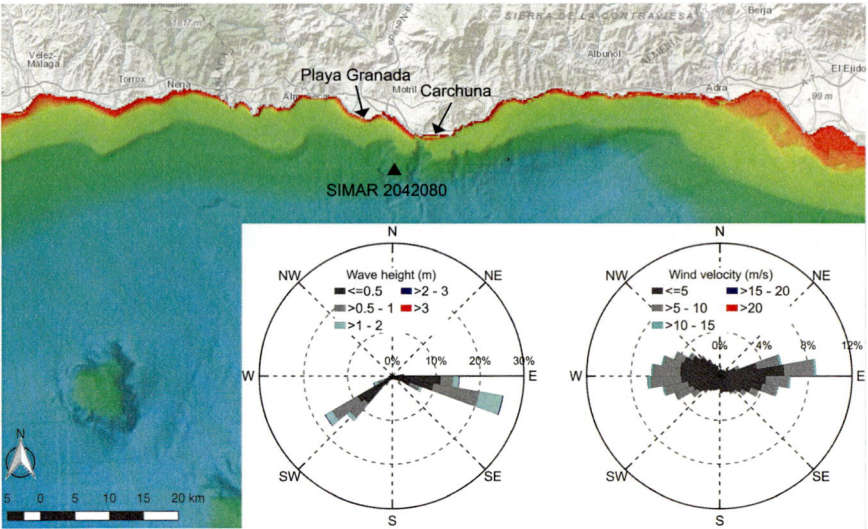

Fig. 2.8 Wave and wind roses at the study sites. Data from SIMAR XXX were provided by *Puertos del Estado*

Acknowledgements This work was partially supported by the Spanish Ministry of Economy and Competitiveness (Projects CTM2012-32439 and BIA2015-65598-P). The second author was funded by the Spanish Ministry of Economy and Competitiveness (Research Contract BES-2013-062617 and Mobility Grant EEBB-I-16-11009). We thank Servicio Provincial de Costas (Granada, Spain) for providing information about the artificial replenishments projects carried out in Playa Granada.

References

1. Ballesteros, M., Rivera, J., Muñoz, A., Muñoz-Martín, A., Acosta, J., Carbó, A., Uchupi, E.: Alboran Basin, Southern Spain-Part II: Neogene tectonic implications for the orogenic float model. Mar. Pet. Geol. **25**(1), 75–101 (2008)
2. Bergillos, R.J., Delgado-Rodríguez, C., López-Ruiz, A., Millares, A., Ortega-Sánchez, M., Losada, M.A.: Recent human-induced coastal changes in the Guadalfeo river deltaic system (southern Spain). In: Proceedings of the 36th IAHR-International Association for Hydro-Environment Engineering and Research World Congress (2015)
3. Bergillos, R.J., López-Ruiz, A., Ortega-Sánchez, M., Masselink, G., Losada, M.A.: Implications of delta retreat on wave propagation and longshore sediment transport—Guadalfeo case study (southern Spain). Mar. Geol. **382**, 1–16 (2016)
4. Bergillos, R.J., Ortega-Sánchez, M., Losada, M.A.: Foreshore evolution of a mixed sand and gravel beach: The case of Playa Granada (Southern Spain). In: The Proceedings of the Coastal Sediments 2015 (2015)
5. Bergillos, R.J., Ortega-Sánchez, M., Masselink, G., Losada, M.A.: Morpho-sedimentary dynamics of a micro-tidal mixed sand and gravel beach, Playa Granada, Southern Spain. Mar. Geol. **379**, 28–38 (2016)

6. Bergillos, R.J., Rodríguez, C., Millares, A., Ortega-Sánchez, M., Losada, M.A.: Impact of river regulation on a Mediterranean delta: assessment of managed versus unmanaged scenarios. Water Resour. Res. **52**, 5132–5148 (2016). doi:10.1002/2015WR018395

7. Bramato, S., Ortega-Sánchez, M., Mans, C., Losada, M.A.: Natural recovery of a mixed sand and gravel beach after a sequence of a short duration storm and moderate sea states. J. Coast. Res. **28**(1), 89–101 (2012)

8. Bárcenas, P., Lobo, F.J., Macías, J., Fernández-Salas, L.M., del Río, V.D.: Spatial variability of surficial sediments on the northern shelf of the Alborán Sea: the effects of hydrodynamic forcing and supply of sediment by rivers. J. Iberian Geol. **37**(2), 195 (2011)

9. Capel-Molina, J.J.: Génesis de las inundaciones de Octubre de 1973 en el Sureste de la Península Ibérica. Cuadernos geográficos de la Universidad de Granada **4**, 149–166 (1974). [in Spanish]

10. Fernández-Salas, L.M., Dabrio, C.J., Goy, J.L., del Río, V.D., Zazo, C., Lobo, F.J., Sanz, J.L., Lario, J.: Land-sea correlation between Late Holocene coastal and infralittoral deposits in the SE Iberian Peninsula (Western Mediterranean). Geomorphology **104**(1), 4–11 (2009)

11. Fernández-Salas, L.M., Lobo, F.J., Sanz, J.L., Diaz-del Rio, V., García, M.C., Moreno, I.: Morphometric analysis and genetic implications of pro-deltaic sea-floor undulations in the Northern Alboran Sea margin, Western Mediterranean Basin. Mar. Geol. **243**(1), 31–56 (2007)

12. Félix, A., Baquerizo, A., Santiago, J.M., Losada, M.A.: Coastal zone management with stochastic multi-criteria analysis. J. Environ. Manage. **112**, 252–266 (2012)

13. Herrero, J., Polo, M.J., Moñino, A., Losada, M.A.: An energy balance snowmelt model in a Mediterranean site. J. Hydrol. **371**(1), 98–107 (2009)

14. Hoffmann, G.: Holozänstratigraphie und Küstenlinienverlagerung an der andalusischen Mittelmeerküste (Doctoral dissertation. Ph.D. thesis, Universität Bremen (1987)

15. Jabaloy-Sánchez, A., Lobo, F., Azor, A., Martin-Rosales, W., Pérez-Peña, J.V., Bárcenas, P., Macias, J., Fernández-Salas, L.M., Vázquez-Vilchez, M.: Six thousand years of coastline evolution in the Guadalfeo deltaic system (Southern Iberian Peninsula). Geomorphology **206**, 374–391 (2014)

16. Lobo, F., Fernandez-Salas, L.M., Moreno, I., Sanz, J.L., Maldonado, A.: The sea-floor morphology of a Mediterranean shelf fed by small rivers, Northern Alboran Sea Margin. Cont. Shelf Res. **26**(20), 2607–2628 (2006)

17. Lobo, F.J., Goff, J.A., Mendes, I., Bárcenas, P., Fernández-Salas, L.M., Martín-Rosales, W., Macias, J., del Río, V.D.: Spatial variability of prodeltaic undulations on the Guadalfeo River prodelta: support to the genetic interpretation as hyperpycnal flow deposits. Mar. Geophys. Res. **36**(4), 309–333 (2015)

18. Losada, M.A., Baquerizo, A., Ortega-Sánchez, M., Avila, A.: Coastal evolution, sea level and assessment of intrinsic uncertainty. J. Coast. Res. **SI59**, 218–228 (2011)

19. López-Ruiz, A., Bergillos, R.J., Ortega-Sánchez, M., Lobo, F.J., Losada, M.A.: Influence of submerged undulations on the development of a horn-embayment system: a case of study in southern Spain. In: Proceedings of the 36th IAHR-International Association for Hydro-Environment Engineering and Research World Congress (2015)

20. Millares, A., Gulliver, Z., Polo, M.J.: Scale effects on the estimation of erosion thresholds through a distributed and physically-based hydrological model. Geomorphology **153**, 115–126 (2012)

21. Millares, A., Polo, M.J., Moñino, A., Herrero, J., Losada, M.A.: Bedload dynamics and associated snowmelt influence in mountainous and semiarid alluvial rivers. Geomorphology **206**, 330–342 (2014)

22. Millares, A., Polo, M.J., Moñino, A., Herrero, J., Losada, M.A.: Reservoir sedimentation and erosion processes in a snow-influenced basin in Southern Spain. In: Reservoir Sedimentation, pp. 91–98. CRC Press (2014)

23. Muñoz, A., Ballesteros, M., Montoya, I., Rivera, J., Acosta, J., Uchupi, E.: Alborán basin, Southern Spain-part I: geomorphology. Mar. Pet. Geol. **25**(1), 59–73 (2008)

24. Ortega-Sánchez, M., Baquerizo, A., Losada, M.A.: On the development of large-scale cuspate features on a semi-reflective beach: Carchuna beach, Southern Spain. Mar. Geol. **198**, 209–223 (2003)

25. Ortega-Sánchez, M., Bramato, S., Quevedo, E., Mans, C., Losada, M.A.: Atmospheric-hydrodynamic coupling in the nearshore. Geophys. Res. Lett. **35**, L23,601 (2008)
26. Ortega-Sánchez, M., Lobo, F.J., López-Ruiz, A., Losada, M.A., Fernández-Salas, L.M.: The influence of shelf-indenting canyons and infralittoral prograding wedges on coastal morphology: the Carchuna system in Southern Spain. Mar. Geol. **347**, 107–122 (2014)
27. Quevedo, E., Baquerizo, A., Losada, M.A., Ortega-Sánchez, M.: Large-scale coastal features generated by atmospheric pulses and associated edge waves. Mar. Geol. **247**, 226–236 (2008)

Chapter 3
Importance of Nearshore Waves on Mixed Sand and Gravel Coasts

Abstract When waves propagate from deep water toward the coast, they modify their properties due to their interaction with the seabed. The Mediterranean basin is characterized by narrow and complex inner shelves that influence the properties of the nearshore waves. This chapter models the wave propagation patterns at the two study sites and highlights the importance of the role played by the inner shelves in the coastal hydrodynamics. A calibrated wave propagation numerical model is applied for that purpose. The results reveal the importance of the feedback process between the forcing (mainly nearshore waves) and the morphological response (changes in the bathymetry) in the evolution of these systems.

3.1 Methodology

3.1.1 Bathymetries

3.1.1.1 Carchuna Beach

As waves propagate from deep water to the coast, they are affected by the seabed when they reach intermediate and shallow water regions. As a result, physical propagation processes, such as shoaling and refraction, are affected [5]. The main elements of the Carchuna shelf that transform wave characteristics, i.e., the canyon, its western and eastern tributaries, and the geomorphological characteristics of the infralittoral prograding wedge (IPW) are plotted in Fig. 3.1.

To analyze the influence of the different bathymetric elements on the coastal morphology, the wave energy distribution under different forcing conditions was analyzed for the actual situation and six synthetic scenarios that were created by removing or smoothing the bathymetric elements, based on a high-resolution bathymetric survey performed in 2010 using an Elac Seabeam 1185 echo sounder. The wave energy distribution was characterized by assessing the propagation coefficient along the study zone. This coefficient is defined as the ratio between the wave height at a given depth and the offshore wave height. The results were then compared to

© The Author(s) 2017
M. Ortega-Sánchez et al., *Morphodynamics of Mediterranean Mixed Sand and Gravel Coasts*, SpringerBriefs in Earth Sciences,
DOI 10.1007/978-3-319-52440-5_3

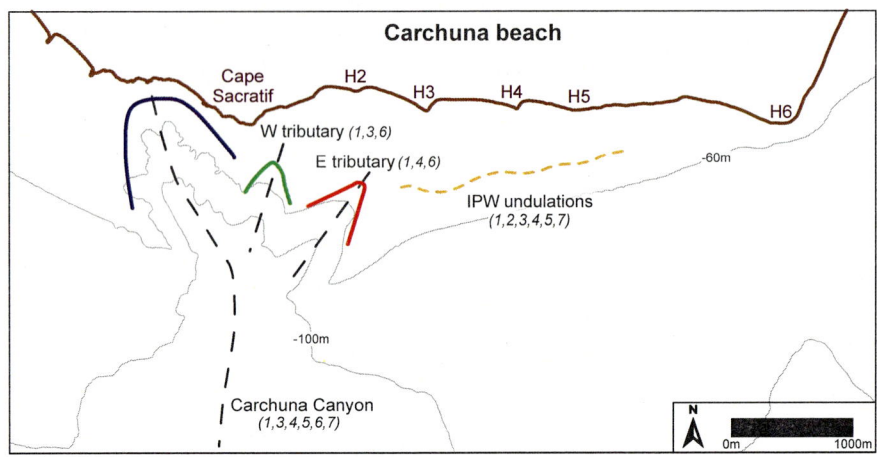

Fig. 3.1 Scheme showing the main bathymetric elements. The *blue line* indicates the Carchuna Canyon. The *green* and *red lines* correspond to the western and eastern tributaries, respectively. The *yellow dashed line* indicates the position of the undulations of the IPW. The *black dashed lines* represent the mean alignment of the Carchuna Canyon and the tributaries. The *numbers* in *parentheses* indicate the scenarios in which the bathymetric element is present. (*Source* [12]. Reproduced with permission of Elsevier)

quantify how the bathymetric elements contribute to the alongshore wave energy modulation.

Scenario 1 corresponds to the actual situation, according to the measured multibeam bathymetric data. In scenario 2, the canyon was eliminated. To achieve this, the original bathymetric contours were split up to remove the canyon and the remaining parts were joined to follow the natural curvature of the shelf. In scenarios 3 and 4, the western and eastern tributaries, respectively, were removed using the procedure described for scenario 2. In scenario 5, both tributaries were removed. In scenario 6, the undulatory pattern exhibited by the IPW offlap break was removed from the bathymetry by straightening the contour lines by smoothing. In scenario 7, both tributaries and the undulatory patterns were removed. A summary of all the scenarios defined is shown in Table 3.1, whereas Fig. 3.2 depicts the most representative scenarios, i.e., scenarios 1, 2, 5, and 6. For the different bathymetric scenarios, the waves were propagated from deep water to the nearshore using the third-generation wave model SWAN (Simulating WAves Nearshore) [2, 7], detailed in Sect. 3.1.2.

3.1.1.2 Playa Granada

High-resolution multibeam bathymetric surveys were carried out in Playa Granada covering the entire deltaic coast in September 1999, October 2004, September 2008 and December 2014 by the Provincial Coastal Service of Granada, the University of Granada, the Spanish Ministry of Environment and Rural and Marine, and the

Table 3.1 Bathymetric scenarios defined for Carchuna Beach

Scenario	Canyon body	Western tributary	Eastern tributary	IPW undulations
1	✓	✓	✓	✓
2	✗	✗	✗	✓
3	✓	✗	✓	✓
4	✓	✓	✗	✓
5	✓	✗	✗	✓
6	✓	✓	✓	✗
7	✓	✗	✗	✗

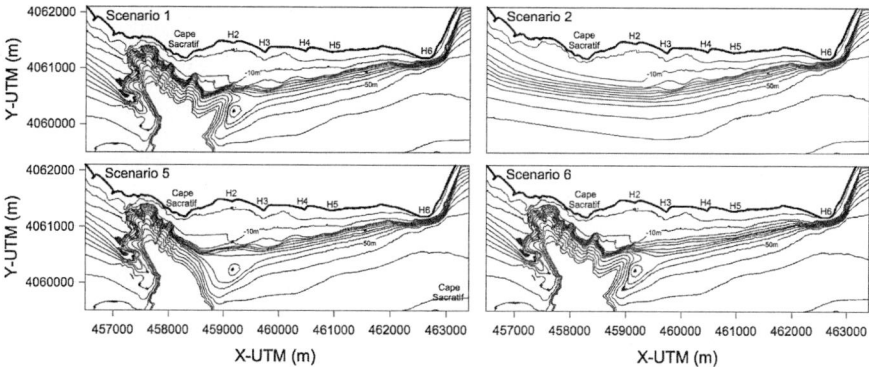

Fig. 3.2 Bathymetries of scenarios 1, 2, 5, and 6 that were used for the wave propagation analyses (isobaths are in meters). (*Source* [12]. Reproduced with permission of Elsevier)

Andalusian Institute for Earth System Research, respectively. The data were acquired using Differential Global Positioning System (DGPS) navigation referring to the WGS-84 ellipsoid. Accurate navigation and real-time pitch, roll and heave were corrected. The multibeam data were also corrected for the sound velocity. These morphological data were used to address the influence of delta retreat and the resulting changes in the submerged morphology (mainly induced by river damming) on the nearshore wave propagation patterns. For that, frequently occurring sea states (Table 3.2), under both low energy and storm conditions, and for both easterly and westerly waves, were propagated from deep water to the nearshore using the WAVE module of the Delft3D model [8, 9]. This module is based on the SWAN model [2, 7], detailed in Sect. 3.1.2.

Table 3.2 Sea states propagated with Delft3D to study the effect of the morphology changes on wave propagation and LST

	Low energy		Storm	
	East	West	East	West
H_0 (m)	0.4	0.4	3.2	3.2
T_p (s)	4.5	4.5	8.4	8.4
θ (°)	112	245	112	245

3.1.2 Wave Propagation Modeling

3.1.2.1 Model Description

The SWAN model was designed to simulate random, short-crested waves in coastal regions [4, 9]. The main processes included in the model are refraction due to bottom and current variations; shoaling, blocking, and reflections due to opposing currents; transmission/blockage through/by obstacles; wind effects; whitecapping; depth-induced wave breaking; bottom friction; and nonlinear wave-wave interactions.

Although refraction is the dominant propagation process [10], to accurately model the surface gravity waves propagating over the complex inner shelf bathymetry, combined refraction and reflection models are required. To this end, [6] demonstrated that, although the SWAN assumptions include small bottom slopes, the alongshore variations of the nearshore wave field caused by refraction over steep shelfs are predicted satisfactorily.

3.1.2.2 Model Implementation: Carchuna Beach

The model domain for Carchuna consists of two different grids, shown in Fig. 3.3. The first grid is a coarse curvilinear 244×73-cell grid covering the entire Carchuna region, with cell sizes that decrease with the depth from 38×55 to 16×24 m. The second grid is a nested grid covering the beach with 325 and 82 cells in the alongshore and cross-shore directions, respectively, and a cell size of 15×15 m. For the spectral resolution of the frequency space, 24 logarithmically distributed frequencies ranging from 0.05 to 1 Hz were used, whereas for the directional space, 72 directions covering 360° in increments of 5° were defined. Simulated data at isobaths of 5 and 8 m were extracted for each scenario for further analysis.

Although the SWAN model was previously applied to Carchuna and validated with data captured during a 7 day field survey [3], we also compared the model results with the data obtained with the instrument moored in 2010 (Fig. 3.3). Figure 3.4 depicts the wave height measured by the instrument during the field survey and the equivalent wave height propagated with the SWAN model for the same location. The

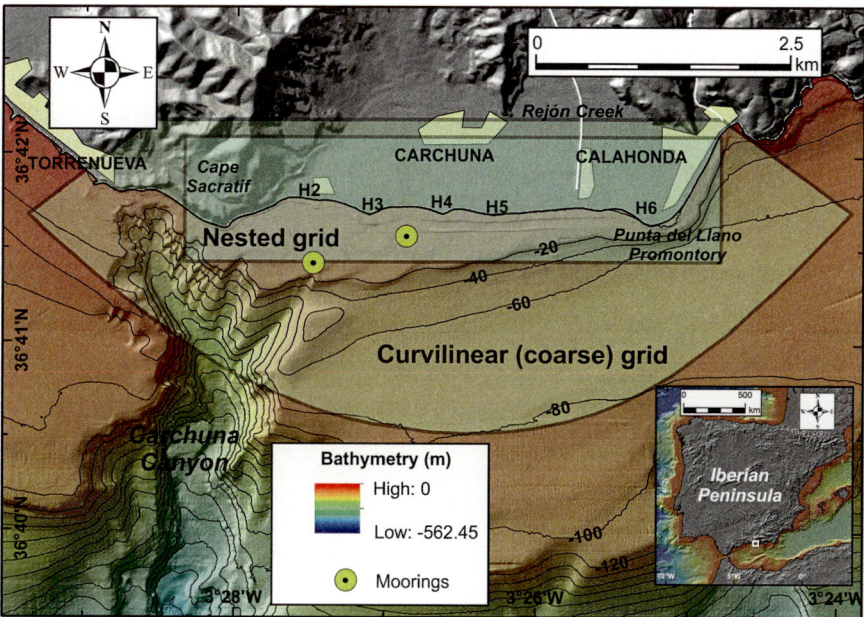

Fig. 3.3 Shaded-relief bathymetry of the study area, comprising the littoral and adjacent sedimentary wedges, indicating the location of the grids (the *curvilinear coarse grid* and the *nested grid*) used in the numerical model and moorings. The bathymetric contours are given in meters below the present sea level. (*Source* [12]. Reproduced with permission of Elsevier)

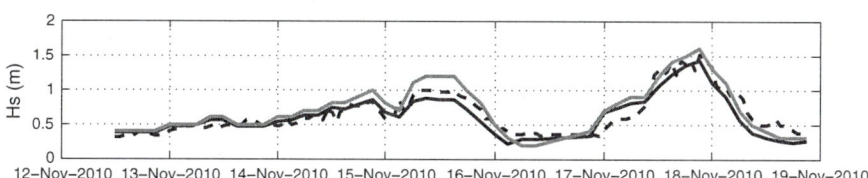

Fig. 3.4 Comparison between the offshore SIMAR input wave data (*gray line*), the propagated wave data measured by the ADCP (*dashed black line*), and the SWAN wave model results obtained for the same location of the ADCP (*black line*). The wave model results correspond to the propagation over the real bathymetry (scenario 1). (*Source* [12]. Reproduced with permission of Elsevier)

wave heights propagated by the model were shown to fit well the measured wave heights.

As stated before, WSW and E are the predominant incoming wave directions. Hence, to describe the general energy distribution on the beach, the following representative incoming wave cases were simulated: WSW and E directions, characteristic low energy and storm wave heights (1 and 4 m, respectively), and short and long peak periods (6 and 10 s), based on the classification defined by [11].

3.1.2.3　Model Implementation: Playa Granada

The model domain consists of two different grids, shown in Fig. 3.5. The first is a coarse curvilinear 82×82-cell grid covering the entire deltaic region, with cell sizes that decrease with decreasing depth from 88×60 to 48×35 m. The second is a nested grid covering the beach with 144 and 82 cells in the alongshore and cross-shore directions, respectively, and cell sizes of about 25×14 m. For the spectral resolution of the frequency space, 37 logarithmically-distributed frequencies ranging from 0.03 to 1 Hz were used; for the directional space, 72 directions covering 360° in increments of 5° were defined.

The model was calibrated through comparison with field data collected from 20 December 2014 to 30 January 2015 by means of two ADCPs (Fig. 3.5). The wave model was forced with the WANA point data (Fig. 3.5) using the bathymetry of 2014 and considering the following physical processes: wind effects, refraction, white-capping, depth-induced breaking ($\alpha = 1$, $\gamma = 0.73$), nonlinear triad interactions ($\alpha = 0.1$, $\beta = 2.2$), bottom friction (Type *Collins*, coefficient = 0.02) and diffraction (smoothing coefficient = 0.6, smoothing steps = 600). Significant wave heights measured by the instruments were compared with the equivalent wave heights propagated with the model for the same locations. Coefficients of determination (R^2)

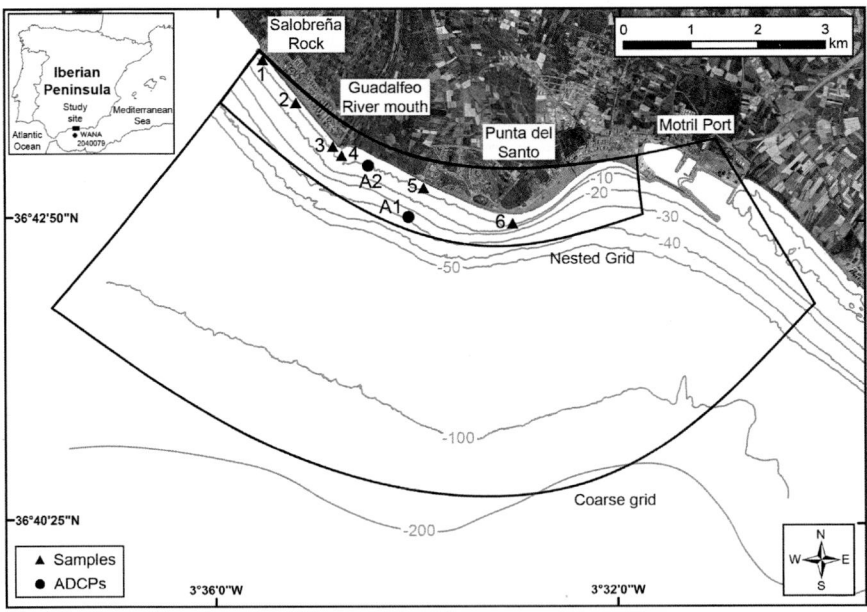

Fig. 3.5 Location and bathymetry of the study site, indicating the profiles where the sediment samples were taken (*red numbered triangles*), the ADCPs (*blue circles A1 and A2*) and the grids used in the numerical model. (*Source* [1]. Reproduced with permission of Elsevier)

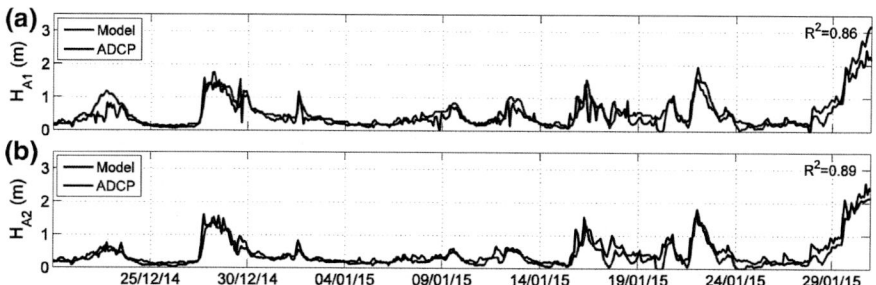

Fig. 3.6 Comparison of time series measured and modelled wave heights in locations A1 (**a**) and A2 (**b**), according to Fig. 3.5. (*Source* [1]. Reproduced with permission of Elsevier)

higher than 0.86 were obtained (Fig. 3.6), providing confidence in the applied wave propagation model.

3.2 Results for the Horn-Embayment System: Carchuna Beach

Based on the analysis of the data obtained at hindcasting point WANA204279, the wave directionality approaches result 47.70% and 45.64% of the time for the E-ESE and W-WSW-SW, respectively, with two main predominant directions: WSW (30.11%) and E (31.66%). The significant wave heights are predominantly less than 1 m (76.9%), whereas storms, that are characterized by significant wave heights greater than 3 m, occur more than once per year on average.

Figure 3.7 shows an example of the SWAN model results for the real bathymetry (scenario 1) under storm conditions ($H_s = 4$ m, $T_p = 10$ s). For both incoming directions (WSW and E), the wave energy is modulated alongshore, with areas of concentration and divergence. The WSW waves concentrate energy between Cape Sacratif and H3 and in the western part of H4. The energy is significantly reduced in the rest of the beach. In contrast, although the results obtained for the E waves showed some areas of wave energy concentration, the energy content was generally significantly lower throughout the beach.

Figure 3.8 represents the alongshore evolution of the propagation coefficient for these two cases. An alongshore modulation pattern is clearly observed, and similar results were obtained for both low energy ($H_s = 1$ m) and storm waves ($H_s = 4$ m). The waves approaching from the WSW are more energetic, with propagation coefficients close to 1 at the western boundary of the beach. These coefficients decrease to the east, but remain consistently greater than 0.6. In contrast, the waves approaching from the E have a coefficient of less than 0.6, with a minimum of 0.3 close to the canyon. The difference between the propagation coefficients in the west and the east

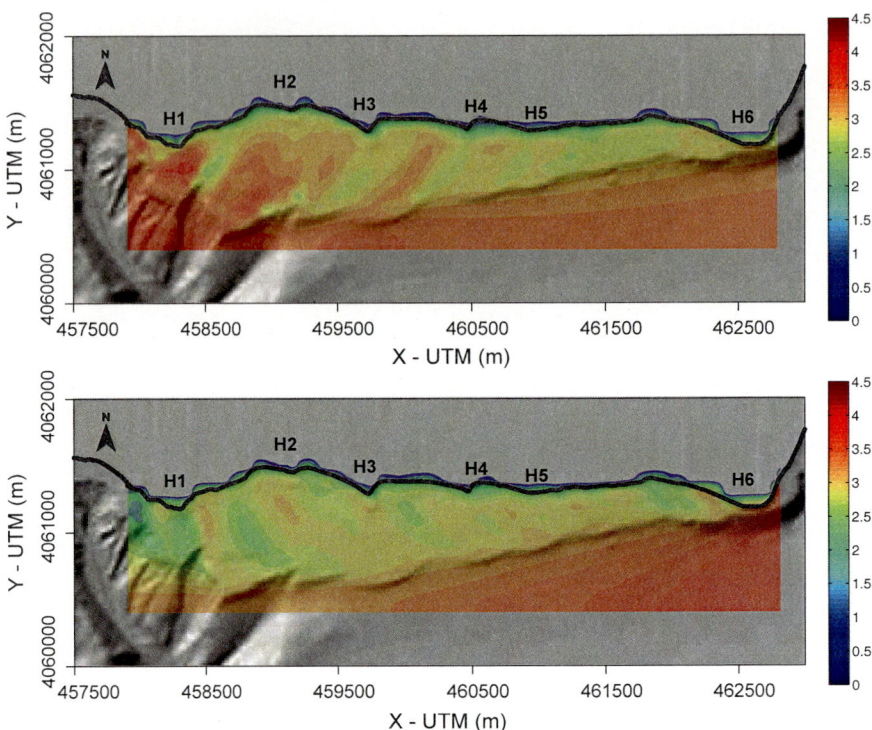

Fig. 3.7 Nearshore wave patterns for the actual bathymetry (sc1). *Upper panel* W incoming waves (from the *left lower corner* of the figure). *Lower panel* E incoming waves (from the *right lower corner* of the figure). Both cases correspond to storm conditions (a significant wave height of 4 m and a peak period of 10 s). (*Source* [12]. Reproduced with permission of Elsevier)

of the study zone shows that the wave heights in the west are approximately 20% higher than in the east of the study zone for the same offshore conditions.

3.2.1   Role of Shelf-Indenting Canyons

Because most of the wave energy modulates and concentrates near the canyon, its influence on the wave propagation was analyzed. The wave propagation results for bathymetric scenario 2 (no canyon) were compared with those obtained for scenario 1 (actual bathymetry). Figure 3.9 shows the difference in the propagation coefficients between the two scenarios for various forcing conditions. Positive (negative) values correspond to energy increases (decreases) due to the canyon. The results show that the canyon concentrates wave energy between H1 and H3, whereas to the east of H3, the nearshore waves behave in a similar manner for both scenarios 1 and 2, with no differences in the propagation coefficients. The presence of the canyon increases the

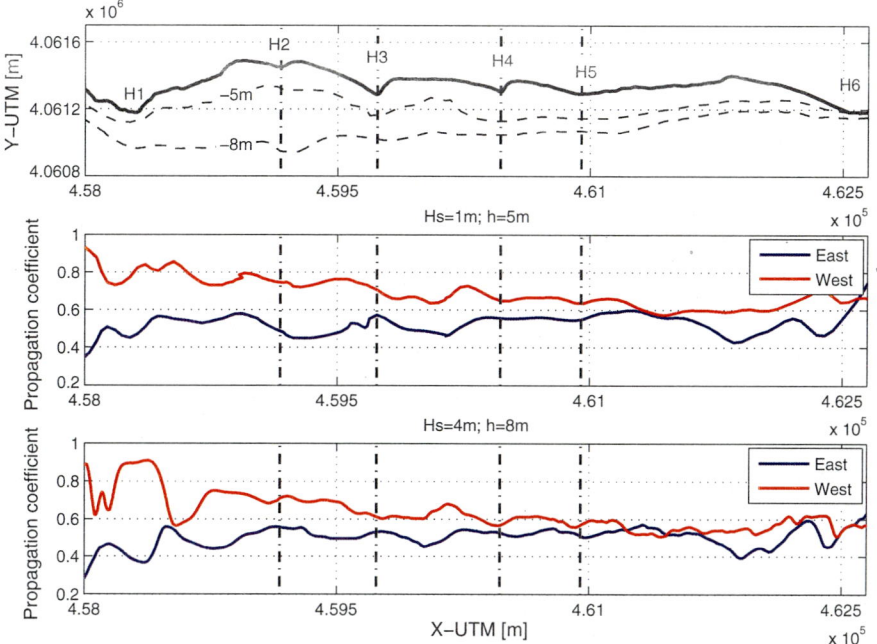

Fig. 3.8 Evolution of the wave energy content for the actual bathymetry (sc1). The results correspond to an average period. The wave data were extracted for depths of 5 and 8 m to avoid the influences of wave breaking and reflection near the shoreline. West refers to WSW waves. (*Source* [12]. Reproduced with permission of Elsevier)

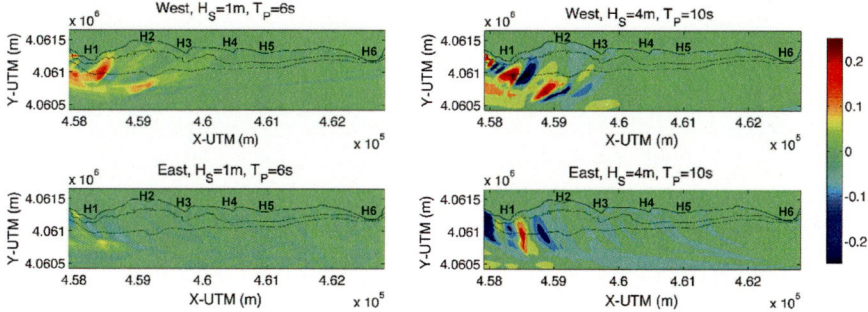

Fig. 3.9 Differences in the propagation coefficients between scenarios 1 and 2. Positive (negative) values indicate increases (decreases) in the wave energy. The *discontinuous contour lines* correspond to the isobaths for depths of 5 and 8 m. West refers to WSW waves. (*Source* [12]. Reproduced with permission of Elsevier)

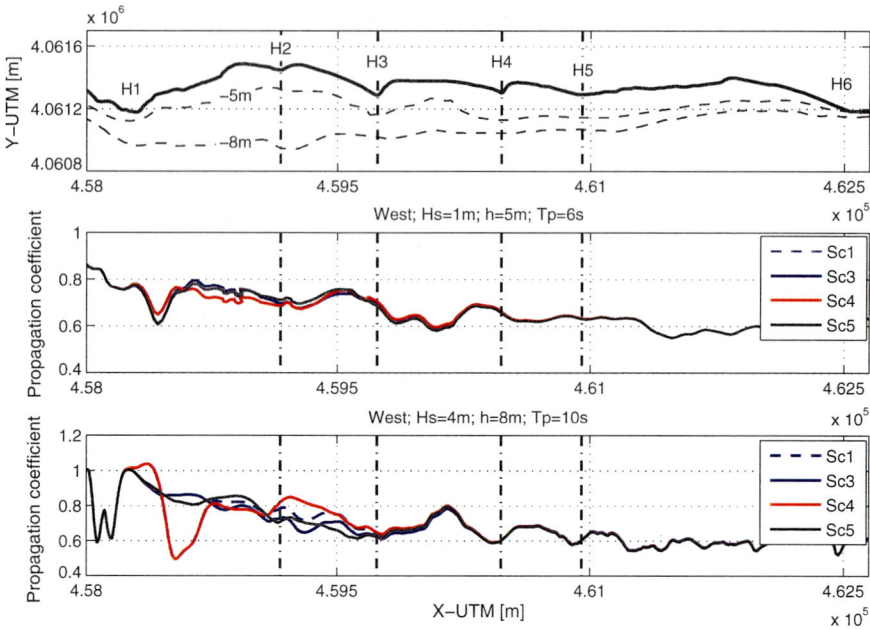

Fig. 3.10 Evolution of the wave energy content for the actual bathymetry (sc1) and the scenarios with the canyon tributaries (sc3, sc4, and sc5). West refers to WSW waves. (*Source* [12]. Reproduced with permission of Elsevier)

nearshore wave heights up to approximately 25%, with a corresponding increase in the wave energy. Because the canyon is located along the western boundary of the beach, its influence is greater on westerly waves than on easterly waves.

Figure 3.10 shows the alongshore evolution of the propagation coefficients for scenarios 1, 3, 4, and 5. The western tributary of the canyon does not exert a significant influence on the wave patterns: the curves for scenarios 1 and 4 are coincident close to H1, and there are significant differences only in front of H2 (more significant at a depth of 8 m than at a depth of 5 m). In contrast, the eastern tributary has a greater influence on the wave modulation: the reduction in the propagation coefficient off Cape Sacratif, which may be responsible for the circulation patterns, disappears. Very similar behavior was observed for the scenario in which both tributaries were eliminated (scenario 5), but some alongshore oscillations were observed. To the east of H3, the influence of the tributaries disappears, and all of the propagation curves collapse into a single one.

3.2.2 *Influence of Infralittoral Prograding Wedges*

Despite the previous studies on the morphology of Carchuna Beach, it is still unclear how the peculiar non-rhythmic large-scale shoreline features are formed. The presence of Carchuna Canyon results in a concentration of wave energy between H1 and H3 (Figs. 3.7 and 3.8), but the canyon does not affect nearshore waves between H3 and H6 for WSW waves (Fig. 3.9), where some of those large-scale cuspate features are found. The possible influence of the submerged IPW undulations was, therefore, analyzed.

Figure 3.11 presents the alongshore evolution of the difference between the propagation coefficients for scenarios 1 and 6 (no submerged undulations) under the predominant wave conditions. For both westerly and easterly waves, an alongshore variation in the wave height on the order of 5–10% was observed. The maximum were found between the horns of the large-scale shoreline features, although higher values were obtained for the western waves than for the eastern waves. At a depth of 8 m, the maximum and minimum for the W and E waves are not at the same

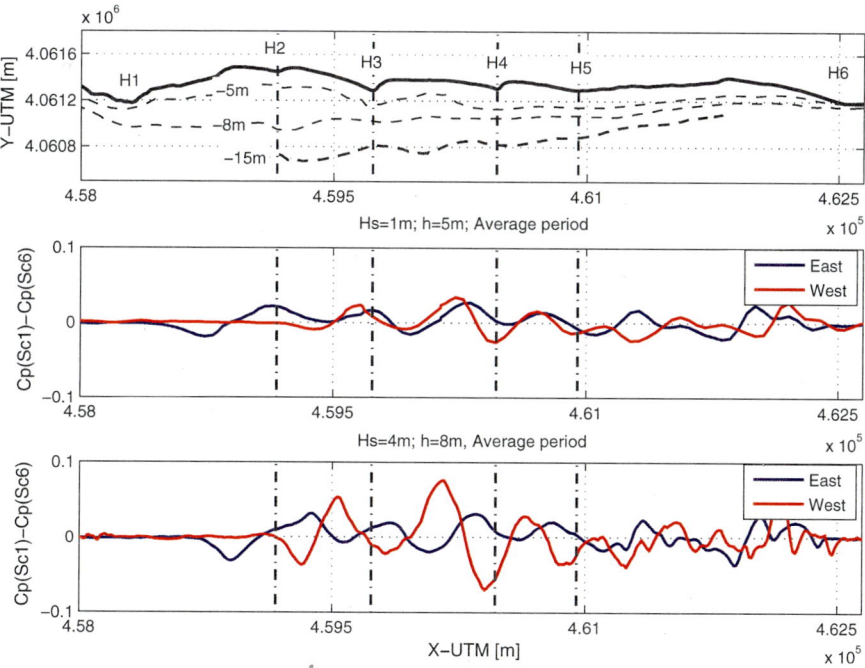

Fig. 3.11 Alongshore evolution of the differences in the propagation coefficients between sc1 and sc6. Positive (negative) values indicate increases (decreases) in the wave height due to undulations. For clarity, the 15 m bathymetric line, where the shape of the undulations is more pronounced, was included in the *upper graph* of the figure. West refers to WSW waves. (*Source* [12]. Reproduced with permission of Elsevier)

locations, whereas at a depth of 5 m, the waves are more refracted and shoaled, and both the amplitude and locations are similar for the W and E waves.

3.3   Implications of Deltas Retreat: Playa Granada

To understand the delta mouth dynamics in terms of nearshore wave propagation patterns, the spatial and temporal variation in the breaking wave height (H_b) around the river mouth was analyzed under low energy and storm conditions (Table 3.2). Figure 3.12 depicts the alongshore distribution of H_b for the four bathymetries detailed in Sect. 3.1.1.2 under both westerly and easterly waves.

Under low energy conditions, the H_b values, and their alongshore variation, are similar for the four bathymetries under both westerly and easterly waves (Fig. 3.12c–e); however, significant differences are observed for typical storm conditions (Fig. 3.12d–f). Under westerly storm waves, it is observed how the delta retreat between 2004 and 2008 (Fig. 3.13b), and the resulting reduction in wave refraction, induced significantly larger values of H_b at the eastern flank of the mouth. The maximum difference represents an increase of almost 10% (difference in H_b of 0.22 m).

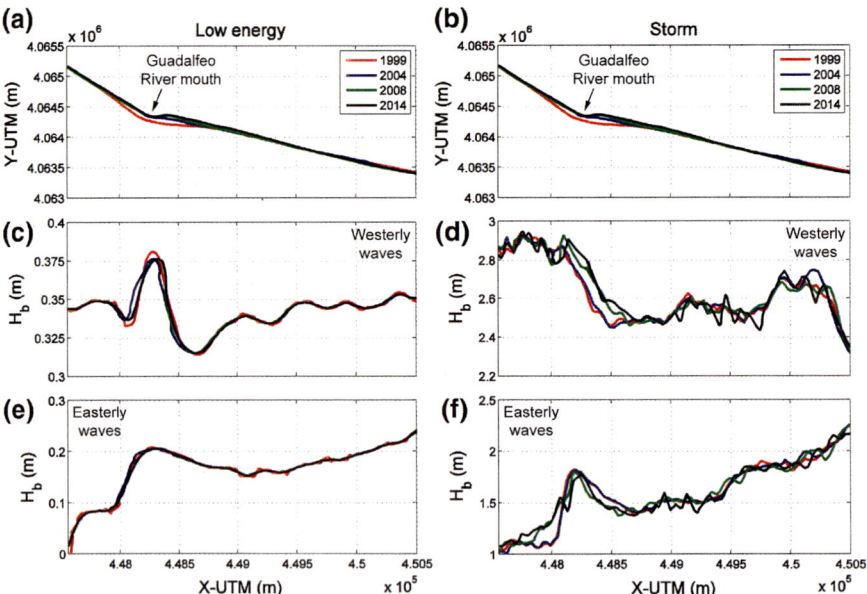

Fig. 3.12 Alongshore evolution of the breaking wave height: low energy conditions for south-westerly (**c**) and easterly waves (**e**); storm conditions for south-westerly waves (**d**) and easterly waves (**f**). The shorelines of the four bathymetries are shown in panels **a** and **b** (*Source* [1]. Reproduced with permission of Elsevier)

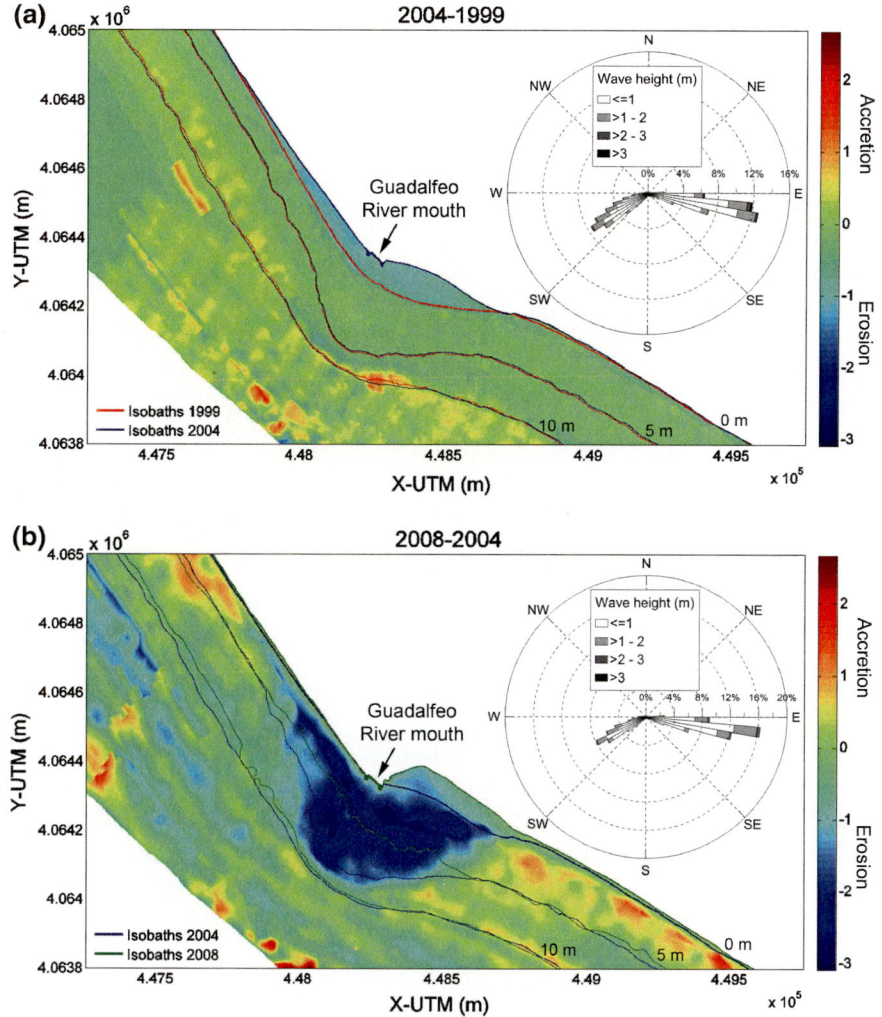

Fig. 3.13 Difference in bed level between multibeam bathymetries and polar diagram showing the frequency of occurrence and the incoming wave directions: **a** 1999–2004, **b** 2004–2008, **c** 2008–2014. (*Source* Adapted from [1]. Reproduced with permission of Elsevier)

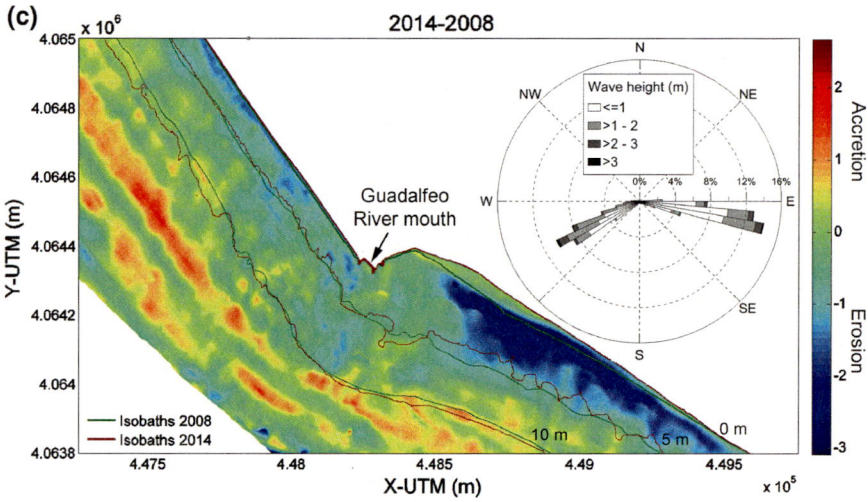

Fig. 3.13 (continued)

Under easterly storm conditions (Fig. 3.12f), higher H_b values are observed in 1999 and 2004 at the eastern flank of the mouth, probably due to the advanced seaward coastline position in 1999 at this location (Fig. 3.13a). However, the delta retreat during the period 2004–2008 led to greater values of H_b at the western flank of the mouth, with differences up to 21% and 0.25 m. During the 2008–2014 period, changes in the wave breaking conditions were significantly less than during the period 2004–2008 (increases of 2% and 10% in H_b for westerly and easterly storm conditions, respectively). This is attributed to the lower erosion rates between 2008 and 2014 at this location (Fig. 3.13c) and the ensuing more stable nearshore morphology in terms of wave propagation.

3.4 Conclusions

1. The wave propagation results along the Carchuna horn-embayment system show that the canyon is the main source of the alongshore modulation of the nearshore waves and that it plays a major role in the generation of the large-scale cuspate features of the shoreline. The presence of the canyon increases the nearshore wave heights by up to approximately 25%. Additionally, the westerly nearshore wave heights are about 20% higher that the easterly ones under the same offshore conditions. Thus, the former dominate the littoral drift.
2. The IPW in Carchuna, and its undulating shape, modified the wave propagation patterns and reinforced the shoreline features. The presence of the IPW results in an alongshore variation in the wave height of the order of 5–10% for both

westerly and easterly predominant waves. When the nearshore waves are modeled considering both the effect of the canyon and the IPW undulations, a strong correspondence between the maximum (minimum) wave heights and the shoreline embayments (horns) was observed.

3. The erosion in delta wedge of Playa Granada since the entry into operation of Rules' Reservoir in 2004 induced lower refraction and greater wave energy around the Guadalfeo River mouth. The breaking wave height under storm conditions increased up to 10% (21%) at the eastern (western) flank of the mouth for incoming westerly (easterly) waves.

Acknowledgements This work was partially supported by the Spanish Ministry of Economy and Competitiveness (Projects CTM2012-32439 and BIA2015-65598-P). The second author was funded by the Spanish Ministry of Economy and Competitiveness (Research Contract BES-2013-062617 and Mobility Grants EEBB-I-15-10002 and EEBB-I-16-11009). The authors are in indebted to Francisco J. Lobo and Gerd Masselink for their valuable suggestions and comments. We also thank Fátima Pereira for her support creating the bathymetric scenarios of Carchuna Beach and Cristóbal Rodríguez-Delgado for his assistance with the bathymetric data processing of Playa Granada.

References

1. Bergillos, R.J., López-Ruiz, A., Ortega-Sánchez, M., Masselink, G., Losada, M.A.: Implications of delta retreat on wave propagation and longshore sediment transport—Guadalfeo case study (southern Spain). Mar. Geol. **382**, 1–16 (2016)
2. Booij, N., Ris, R.C., Holthuijsen, L.H.: A third-generation wave model for coastal regions: 1. Model description and validation. J. Geophys. Res. Oceans **104**(C4), 7649–7666 (1999)
3. Bramato, S., Ortega-Sánchez, M., Mans, C., Losada, M.A.: Natural recovery of a mixed sand and gravel beach after a sequence of a short duration storm and moderate sea states. J. Coast. Res. **28**(1), 89–101 (2012)
4. Dan, S., Djr, W., Stive, M.J., Panin, N.: Processes controlling the development of a river mouth spit. Mar. Geol. **280**(1), 116–129 (2011)
5. Dean, R.G., Dalrymple, R.A.: Coastal Processes with Engineering Applications. Cambridge University Press (2002)
6. Gorrell, L., Raubenheimer, B., Elgar, S., Guza, R.T.: SWAN predictions of waves observed in shallow water onshore of complex bathymetry. Coast. Eng. **58**, 510–516 (2011)
7. Holthuijsen, L.H., Booij, N., Ris, R.C.: A spectral wave model for the coastal zone. In: Ocean Wave Measurement and Analysis, pp. 630–641. ASCE (1993)
8. Lesser, G.R.: An Approach to Medium-Term Coastal Morphological Modelling. UNESCO-IHE, Institute for Water Education (2009)
9. Lesser, G.R., Roelvink, J.A., Jatm, V.K., Stelling, G.S.: Development and validation of a three-dimensional morphological model. Coast. Eng. **51**(8), 883–915 (2004)
10. Magne, R., Belibassakis, K.A., Herbers, T.H., Ardhuin, F., O'Reilly, W.C., Rey, V.: Evolution of surface gravity waves over a submarine canyon. J. Geophys. Res. Oceans **112**, C01,002 (2007). doi:10.1029/2005JC003035
11. Ortega-Sánchez, M., Baquerizo, A., Losada, M.A.: On the development of large-scale cuspate features on a semi-reflective beach: Carchuna beach, Southern Spain. Mar. Geol. **198**, 209–223 (2003)
12. Ortega-Sánchez, M., Lobo, F.J., López-Ruiz, A., Losada, M.A., Fernández-Salas, L.M.: The influence of shelf-indenting canyons and infralittoral prograding wedges on coastal morphology: the Carchuna system in Southern Spain. Mar. Geol. **347**, 107–122 (2014)

Chapter 4
Littoral Drift and Coastline Evolution on Mixed Sand and Gravel Coasts

Abstract The evolution of the coastline is mainly driven by the gradients in the longshore sediment transport (LST). These gradients are mostly influenced by the incoming waves and the morphology of the coastline. Traditionally, the coastline is assumed to be quasi-rectilinear, and the formulations are not valid for curved shorelines. This chapter presents a new expression that not only accounts for the curvature of the shoreline but also includes the variability of the sediment size that is typically found along Mediterranean coasts. It is later applied to the study sites described in Chap. 2, and the relation between LST trends and coastline evolution is analyzed and discussed for both sites.

4.1 Longshore Sediment Transport (LST)

4.1.1 LST Expressions

When waves approach the coast, they shoal, refract and break, generally at an oblique angle. During this process, longshore currents are generated due to the cross-shore gradient in the longshore shear component of the radiation stress [23]. Longshore currents may transport a considerable amount of sediment along the coast [12], even for gravel and shingle beaches. This longshore sediment transport (LST hereafter) is a key process in coastal morphodynamics because its alongshore gradients drive changes in the nearshore morphology [11] at temporal scales ranging from hours to centuries and at spatial scales ranging from ten of meters to hundreds of kilometers [22]. These changes are occasionally caused by episodic and large LST rates or persistent longshore LST gradients that drive inlet closures [36], headland bypassing of large amount of sediments [39], rotation of pocket beaches [15], and shoreline recession or persistent erosion on the shorefaces of deltas [7]. Most of these impacts are generally considered to be negative by coastal managers and planners, and over the last several decades, there has been a sound amount of research work focused on developing robust and easy-to-use expressions for LST prediction [29].

© The Author(s) 2017
M. Ortega-Sánchez et al., *Morphodynamics of Mediterranean Mixed Sand and Gravel Coasts*, SpringerBriefs in Earth Sciences,
DOI 10.1007/978-3-319-52440-5_4

Among the numerous works on LST expressions, two different approaches have been followed:

- Process-based approaches. They generally include a more precise description of the nearshore processes, such as wave breaking or wave-current interactions [29]. Although very accurate results can be obtained with this approach, they are highly inefficient because of the high number of input variables that must be prescribed and the need for calibration prior to every application. Two examples of this type of approach are [14, 17].
- Bulk transport expressions based on simplifications of the nearshore physical processes that drive the LST [11]. These types of approaches are the most widely used and represent the best alternative to make first estimates or tendency analyses, especially when the available information is limited.

In this work, we focused on the latter type of approach, which has received much more attention from researchers in recent decades. The first important expression developed was that by [18], which assumes that wave energy is expended to suspend and support the sediment above the bottom. The longshore current, which is superimposed on the orbital velocity of the particles, is responsible for transport of the sediment particles and produces a net drift in the direction of the current. The expression by [18] reads:

$$Q = \frac{K_{\text{I\&B}}}{(\rho_s - \rho)g(1 - p)} C_{gb} E_b \cos \alpha_b \left(\frac{V}{u_0} \right) \qquad (4.1)$$

where Q is the volumetric LST rate in m³/s, ρ and ρ_s are the water and particles densities, respectively, g is the acceleration of gravity, p is the porosity index ($\simeq 0.4$), C_{gb} is the group celerity of waves at breaking, E_b is the breaking wave energy, α_b is the wave angle at breaking, V is the longshore current velocity (usually measured at the middle of the surf zone) and u_0 is the maximum horizontal bottom orbital velocity of the waves at the breaker zone. The coefficient $K_{\text{I\&B}}$ is dimensionless and was fitted by [21] using field data, obtaining $K_{\text{I\&B}} = 0.25$.

Two decades later, the CERC equation [40] was developed and remains the most widely used formula for the total LST rate in Coastal Engineering practice. This equation assumes that the LST is proportional to the available longshore wave power per unit of beach length and is expressed as:

$$Q = K_{\text{CERC}} \frac{\rho \sqrt{g}}{16\sqrt{\gamma}(\rho_s - \rho)(1 - p)} H_{sb}^{5/2} \sin (2\alpha_b) \qquad (4.2)$$

where γ is the breaking index and H_{sb} is the significant wave height at breaking. The coefficient K_{CERC} is dimensionless and empirical, and different values have been proposed. The Shore Protection Manual [40] recommended $K = 0.39$, whereas [38] experimentally fitted a value of $K = 0.2$. del Valle et al. [41] found a relation between this coefficient and the median grain diameter of the sediment (D_{50}):

$$K_{\text{CERC}} = 1.4 \exp\left(-2.5 D_{50}\right) \tag{4.3}$$

Kamphuis [19] defined an expression for the LST rate based on physical model experiments that was later generalized for both field and laboratory purposes [20]. It was based on a dimensional analysis along with some physical assumptions, and its main advantages are that it includes the wave period, beach slope, and grain size. The simplified version of the [19] expression is

$$Q = \frac{2.27}{(\rho_s - \rho)(1 - p)} H_{sb}^2 T_p^{1.5} \tan \beta^{0.75} D_{50}^{-0.25} \sin (2\alpha_b)^{0.6} \tag{4.4}$$

where T_p is the peak wave period and $\tan \beta$ is the beach slope in the surf zone.

More recently, [5] developed an LST rate formula based on a hypothesis similar to that of [18]: the sediment becomes suspended by the action of breaking waves and is then transported by longshore currents that can be generated by different agents such as waves or tides. Moreover, this expression assumes that most of the sediment remains in suspension. Then, the work needed to maintain the movement of the sediment particles is a fraction of the wave energy flux and depends on the sediment fall velocity. The final expression is

$$Q = \frac{K_{\text{Bayram}}}{(\rho_s - \rho)g(1 - p)w_s} E_b C_{gb} \overline{V} \tag{4.5}$$

where w_s is the sediment fall velocity, $\overline{V}$ is the mean alongshore current on the surf zone (due to waves, tides or any other agent) and K_{Bayram} is a dimensionless coefficient that accounts for the efficiency of waves in keeping the sediment in suspension, as found through dimensional analysis:

$$K_{\text{Bayram}} = \left(9 + 4\frac{H_{sb}}{w_s T_p}\right) \tag{4.6}$$

Notice that all of the expressions described above include a dimensionless coefficient whose values are fitted or calibrated using different datasets.

4.1.2 Limitations of the LST Expressions for Mixed Coasts

Mixed sand and gravel beaches in the westerly coasts of the Mediterranean Sea are usually characterized by curvilinear shorelines (Fig. 4.1). As demonstrated by [24, 26], these geometries induce important longshore gradients in two main variables related to shoreline morphodynamics, i.e., the wave angle in the surf zone and the wave energy content, that in turn promote the presence of important longshore variations in both grain size [28] and beach slope [11].

Fig. 4.1 Examples of curvilinear coasts in the Spanish coast of the Mediterranean Sea: **a** Playa Granada, **b** Carchuna and **c** La Rábita beaches. *White lines* indicate the curvature of the shoreline (Aerial images courtesy of Apple Inc.)

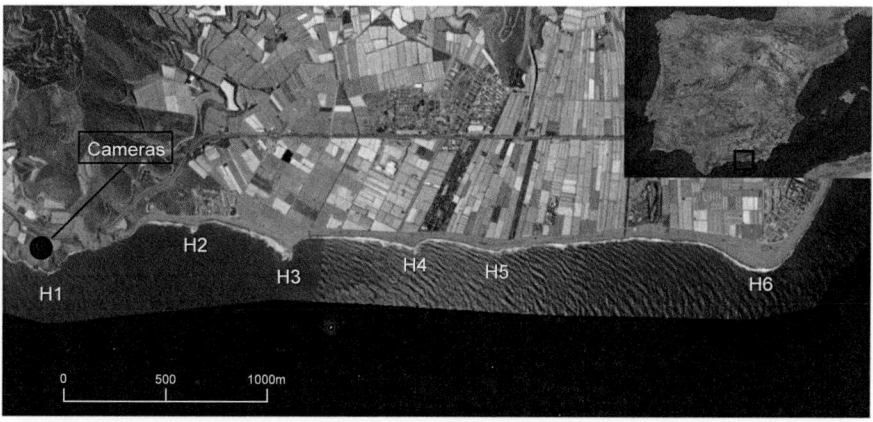

Fig. 4.2 Carchuna Beach system: H1-H6 denote the horns of the large-scale shoreline features and the *blue circle* represents the location of the video monitoring station (Aerial images courtesy of Apple Inc.)

Evidence in nature can be found for both processes. As described in Chap. 2, Carchuna Beach is a mixed coast in southeastern Spain that faces the Alborán Sea. The beach exhibits a series of large-scale cuspate features with an alongshore spacing of hundreds of meters. These features are bounded by a series of seaward-extending horns (H1–H6, Fig. 4.2) that form a curvilinear coast. The beach is bounded to the west by Cape Sacratif (H1) and to the east by the Punta del Llano Promontory (H6), which corresponds to the west side of the Calahonda Spit. Moreover, the bathymetry of the beach is characterized by the presence of submerged morphologies that affect wave propagation (see Chap. 3). The wave climate is characterized by two main wave incoming directions (E and W) that cause waves to reach the nearshore with a significant obliquity.

The Environmental Fluid Dynamics Research Group (University of Granada, Spain) installed a video-monitoring station in November 2002 based on the ARGUS technique [1] at the Sacratif lighthouse (located at H1), 50 m above mean sea level. The station includes 3 video cameras that collect images during the first 10 min of each daylight hour at a frequency of 2 Hz. Video images consist of an instantaneous image (snapshot), a 10 min time-averaged image (timex) and a variance image for the same period. The georeferenced digitization of the beach was performed using the methodology of [16] at one-pixel accuracy. At the midbeach, one pixel corresponds to a ground accuracy of 0.25 and 1.4 m in the cross-shore and alongshore directions, respectively, and worsens to 0.49 and 5.59 m at the far end of the beach.

The images taken by the video-monitoring station were analyzed to check the presence of alongshore variations in both surf zone width, which visually characterized the wave energy content, and nearshore angle. The effort was focused on storm events because the variations in wave propagation are highlighted under these conditions [33].

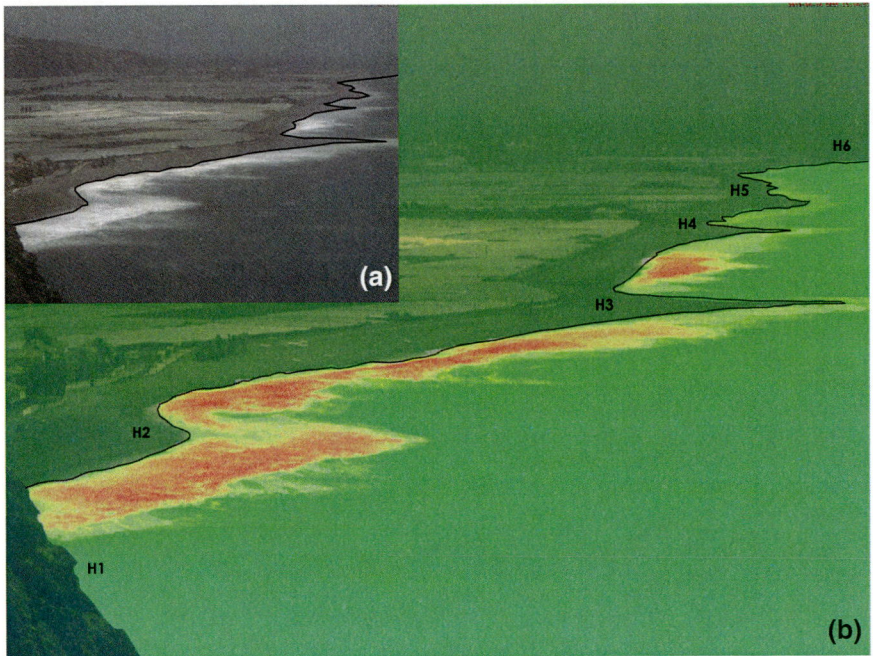

Fig. 4.3 Snapshot of Carchuna Beach on March 14th, 2012 at 13 h: **a** Non-filtered image; **b** Filtered image, where *yellow* and *red* colors identified the surf zone

After the analysis, the presence of alongshore gradients in surf zone width were evident. The wave energy concentrated and diverged in some locations of the beach, with differences in surf zone width of approximately 50% in some cases. The places where the wave energy concentrated were not only the horns, as might be expected. There were also energy concentrations in the embayments, presumably due to the effect of refraction over the curvilinear bathymetry. Moreover, the surf zone width tended to be larger in the side of the horn oriented to the direction where the waves came from.

Figure 4.3 shows an example of these effects. A snapshot of the beach during a westerly storm on March 14, 2012, at 13:00 (Fig. 4.3a) and a filtered image of the snapshot (Fig. 4.3b) are shown. The filter was used to contrast the surf zone (yellow to red colors) over the rest of the image (green colors). The color scale indicates that energy concentrated around the horn H2 and in the embayments between H2-H3 and H3-H4. The patterns in the nearshore wave energy were described and numerically checked by [24] and are summarized in Fig. 4.4.

To determine whether similar effects could be observed for the wave angle, rectified snapshots of the nearshore were analyzed for storm conditions. With this type of image, the relative orientation between the wave crests and the shoreline can be measured. Figure 4.5 shows a rectified plan view of the track of the beach between

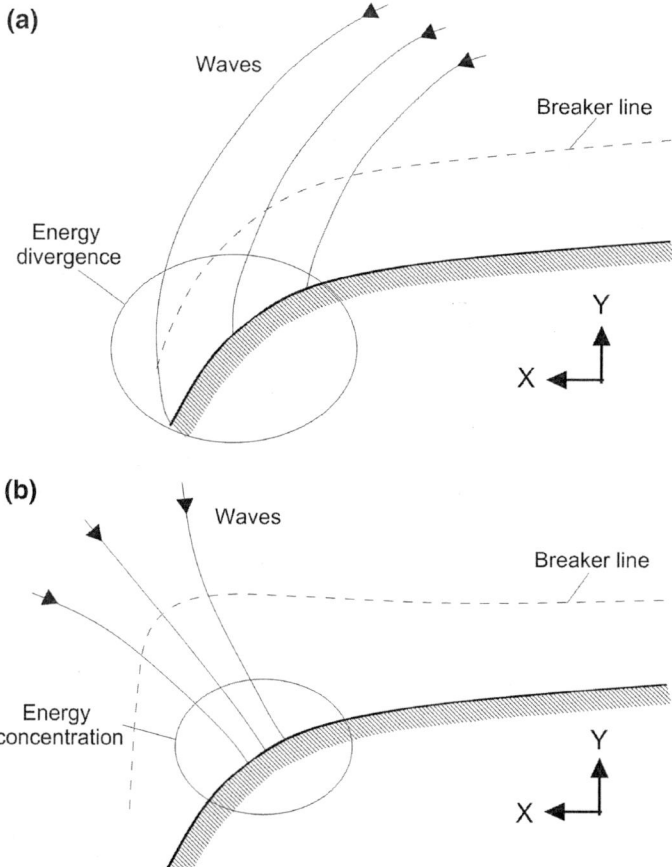

Fig. 4.4 Wave energy pattern at curvilinear coasts: **a** energy divergence associated with high-angle wave approach; **b** energy concentration associated with normal to negative wave angle values. The angles of incidence are measured counterclock- wise from the positive Y-axis. (*Source* [24]. Reproduced with permission of Elsevier)

H3 and H4 on February 2, 2004, at 8h, during an easterly storm. The image shows that the wave fronts were parallel to the shoreline on the left side of the horns, which were the sides of the horns more exposed to waves. However, for the embayments and especially for the right side of the horns, the wave fronts and the shoreline formed a significant angle (approximately 45°).

These results indicate that the presence of a curvilinear coast induces alongshore gradients in both the surf zone width (and hence in wave energy) and nearshore wave angle that limits the applicability of the more widely used LST formulations, such as the CERC [40] or the Kamphuis [19] expressions. The former was obtained considering uniform stretches of coast in the longshore direction, without variations in the beach and sediment characteristics, whereas the latter was obtained empirically

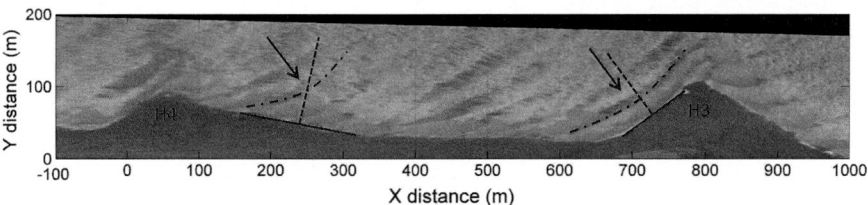

Fig. 4.5 Rectified snapshot of the H3-H4 embayment (Carchuna Beach) for February 2nd, 2004 at 8 h. *Solid line*: local beach alignment; *dashed lines*: normal to shore direction; *dash-dotted lines*: wave front; arrows: wave direction

for sandy beaches with sediment sizes different from those of mixed sand and gravel beaches. Moreover, the [19] expression over-predicts the LST rates on coarse-grained beaches [37].

4.1.3 An Updated Expression of LST for Nonuniform Mixed Coasts

Considering all these limitations, a significant effort was carried out to obtain an alternative LST approach in recent years [24–26]. In this chapter, the energetic approach proposed by [25] is presented and applied. This expression uses a framework specifically defined for curvilinear coasts, considering longshore variations of the shoreline and wave angles, and also gradients in the wave energy characterized by the surf zone width. Moreover, it was obtained without the use of hypothesis of uniform nor regular beach characteristics in the longshore direction. As well as the others energetic approaches, the expression accounts for the potential LST that can be transported given specific wave energy conditions. This potential LST fits the actual LST rate if there is enough sediment of the characteristic grain size to be mobilized. Furthermore, there is another source of uncertainty related with grain sizes, since the use of a characteristic D_{50} implies that the sediment is considered as uniform for a specific beach profile, although it can vary longshore.

The expression by [25], which framework and variables are defined in Fig. 4.6, is an updated version of the LST formula of [18] adapted for curvilinear coasts. This latter expression relates the volume of alongshore sediment transport including pores to the wave energy dissipation along the coast, and can be written as follows:

$$S_{I\&B} = \frac{K}{(\rho_s - \rho)g(1 - p)}(EC_g)_b \cos\theta_b \frac{\bar{V}}{u_{mb}} \tag{4.7}$$

where ρ and ρ_s, are the water and sediment particle densities, p is the porosity of the material, K is a dimensionless coefficient which depends on the grain size [41], u_{mb} is the maximum horizontal bottom orbital velocity of the waves evaluated at the

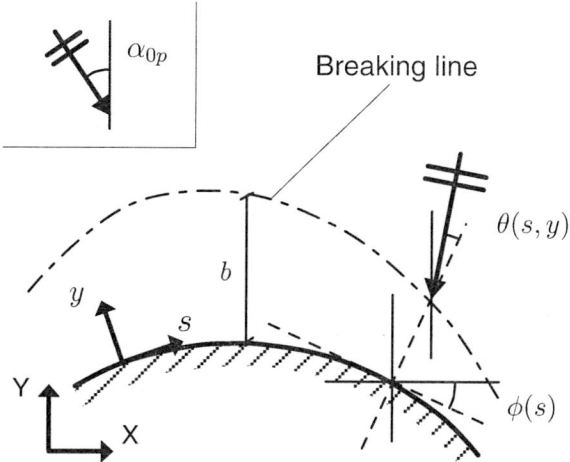

Fig. 4.6 Definition of the main variables as a function of the local coordinates s and y, where α_{0p} is the deep water wave angle respect to the global coordinates X, Y, $\theta(s, y)$ is the local nearshore wave angle, $b(s)$ is the distance between the shoreline and the breaking line in the y direction (surf zone width), and $\phi(s)$ is the local shoreline angle. *Solid lines* show the (X, Y) directions, whereas the *dashed lines* show the local (s, y) directions. (*Source* [25]. Reproduced with permission of Elsevier)

breaking, and $\bar{V}$ is the averaged alongshore current caused by the oblique incidence of the waves in the surf zone [23]. Note that $E_b C_{gb}$ represents the wave energy flux per unit of wave crest that is dissipated in the whole cross-shore section of the surf zone.

The objective of the update LST expression is to include the effects of spatial variations in the sediment size, beach slope, type of wave breaking, wave angle and wave energy. If the wave angle in the nearshore is known, EC_g, V and u_m can be expressed in terms of the local variables s and y (Fig. 4.6). In this case, $E(s, y)C_g(s, y)\cos\theta(s, y)$ represents the wave energy flux per unit of wave crest in the direction of the waves. If nearshore values are retained, the local alongshore current becomes:

$$V(s, y) = \frac{5}{8}\pi g \frac{\tan\beta(s)^2\gamma}{f}\frac{\sin\theta(s, y)}{\sqrt{g\tan\beta(s)y}}y \qquad (4.8)$$

where f is a friction coefficient and γ is the breaking index. Then, linear wave theory and depth-limited breaking can be used to obtain the local alongshore sediment transport per unit of cross-shore section:

$$S_{lst}(s, y) = K_T y^{5/2} \sin(2\theta(s, y))dy \qquad (4.9)$$

where:

$$K_T(s) = \frac{5\,K\pi\rho\,\gamma^2(\tan\beta(s))^{7/2}\sqrt{g}}{32(\rho_s - \rho)(1 - p)f} \tag{4.10}$$

The averaged sediment transport rate parallel to the coastline can then be obtained by integrating this local alongshore sediment transport per unit of shoreline along the surf zone [24]:

$$Q(s) = \frac{1}{b(s)}\int_0^{b(s)} S_{lst}(s, y)dy = K_T(s)\frac{1}{b(s)}\int_0^{b(s)} \sin(2\theta(s, y))y^{5/2}dy \tag{4.11}$$

The integral in Eq. (4.11) has analytical solutions if an analytical expression is adopted for $\theta(s, y)$. Moreover, Eq. (4.11) was obtained without any restriction on the wave angle, the sediment size or the beach slope. Thus, these parameters can be considered variables in both the s and y directions.

4.1.4 A Specific Solution for the LST Expression

In this section, a specific solution of Eq. (4.11) is obtained by means of a simple model for wave propagation to calculate the surf zone width b and the wave angle θ. Although a simple propagation model was chosen, any other solution for wave propagation could be used.

4.1.4.1 Surf Zone Wave Angle

The wave propagation model assumes that if inverse shoaling is considered, the wave ray in deep water has the orientation of the approximation defined. For every shoreline position s, the modified deep-water wave angle is defined as follows:

$$\alpha_0(s) = \alpha_{0p} + \phi(s) \tag{4.12}$$

Hence, every section of the shoreline is considered locally as a rectilinear stretch of the coast with shore-parallel contours. An expression for $\theta(s, y)$ can then be obtained by using Snell's law for shore-parallel contours:

$$k_0 \sin\alpha_0(s) = k \sin\theta(s, y) \tag{4.13}$$

where k is the wave number and the subscript 0 indicates deep-water values. The wave number is obtained by means of linear theory, with $k_0 = \sigma^2/g$ and $k =$

$\sigma/\sqrt{g \tan\beta(s) y}$, $\sigma = 2\pi/T$ being the wave frequency and T the wave period. The wave angle in the nearshore zone yields the following expression:

$$\theta(s, y) = \arcsin\left[\sigma\sqrt{\frac{\tan\beta}{g}} \sin\alpha_0(s)\sqrt{y}\right] = \arcsin\left(\Psi(s)\sqrt{y}\right) \qquad (4.14)$$

where:

$$\Psi(s) = \frac{2\pi}{T}\sqrt{\frac{\tan\beta}{g}} \sin\left(\alpha_{0p} + \phi(s)\right) \qquad (4.15)$$

This parameter can be interpreted as a measure of the wave refraction for a given coast configuration ($\phi(s)$ and $\tan\beta$) or for a given deep-water wave conditions (α_{0p} and T).

The validation of this approximation using the SWAN model [8] is presented in [27]. The approximation differs slightly from the model but is much simpler to implement. Only under highly oblique waves ($\alpha_{0p} > 45°$) the correlations and skills were $R < 0.94$ and $S < 0.6$, respectively. However, the alongshore patterns of wave angle gradients are well reproduced. These gradients drive the alongshore changes in sediment transport, and hence, the coastline evolution [11].

4.1.4.2 Surf Zone Wave Height

To obtain the surf zone width b, the wave height is obtained using the shoaling and refraction coefficients. The former is obtained using the relation between the group celerity in deep water and in the nearshore. The latter is obtained using the conservation equation of wave energy flux along the wave crests. The wave height in the nearshore can then be estimated as follows:

$$H = H_0 K_p = H_0 K_s K_r = H_0\sqrt{\frac{C_{g0}}{C_g}} \sqrt{\frac{d_0}{d}} \qquad (4.16)$$

where d is the distance between wave rays, C_g is the group celerity, the subscript 0 indicates deep water, and K_p, K_s and K_r are the propagation, shoaling and refraction coefficients, respectively. The distance between wave rays can be assessed geometrically or can be calculated solving the ODE system defined by [30] (for completeness, see [25]). The surf zone width b can then be determined using a variable breaking index. This approximation was successfully applied to various wave and beach conditions, considering not very sharp shorelines and waves that are not highly oblique ($\alpha_{0p} < 60°$).

4.1.4.3 Specific Solution for LST

A simple analytical solution of Eq. (4.11) is obtained considering $\tan \beta$ and the sediment size constant in the y direction. Combining Eqs. (4.14) and (4.11), it follows for the averaged sediment transport rate parallel to the coastline:

$$Q(s) = K_T \frac{4}{315b\Psi^7} \left[16 - \left(1 - b\Psi^2\right)^{3/2} \left[16 + b\Psi^2 \left(24 + 5b\Psi^2(6 + 7b\Psi^2)\right)\right]\right]$$
(4.17)

where $Q(s)$ is expressed as a function of the alongshore distribution of beach slope and sediment size (included in K_T and Ψ), the surf zone width, the wave period, and the deep-water wave and shoreline angles (included in Ψ).

Equation (4.11) only has an analytical solution (Eq. 4.17) for combinations of b and Ψ that fulfil breaking depths $h_b \leq gT^2/40$, which is true for the majority of typical wind-generated waves (wave periods between 3 and 30 s). Although Eq. (4.17) is not mathematically defined for $\Psi = 0$ (shore-normal wave incidence), the limit of this function has a finite value: $\lim_{\Psi(s) \to 0} Q(s) = 0$. This implies vanishing alongshore sediment transport for both waves arriving normal to the shoreline and for no incident wave energy. To simplify the application of Eq. (4.17) in the vicinity of $\Psi = 0$, it can be expressed as a power series expansion around these values. The result for the series in the vicinity of $\Psi = 0$ to the order $\mathcal{O}[\Psi^7 b^6]$ is as follows:

$$Q(s) = K_T \left(\frac{\Psi b^3}{2} - \frac{\Psi^3 b^4}{5} - \frac{\Psi^5 b^5}{24} + \mathcal{O}\left[\Psi^7 b^6\right]\right)$$
(4.18)

Considering typical values of $\Psi[\mathrm{m}^{-1/2}] = \mathcal{O}[10^{-2}]$ and $b[\mathrm{m}] = \mathcal{O}[10^2]$, the terms on the right side of this expression decrease as their exponents increase, and the second and subsequent terms on the right side can be neglected with respect to the first term. Thus, for those cases:

$$Q(s) \simeq K_T \left(\frac{b^3 \Psi}{2}\right)$$
(4.19)

which is an expression that is simple to implement in practical applications and that avoids numerical problems in the vicinity of $\Psi = 0$.

4.2 Coastline Evolution Due to Submerged Morphologies: The Case of Carchuna Beach

As described in Chaps. 2 and 3, Carchuna Beach is characterized by a coastline morphology with non-rhythmic large-scale alongshore features (H2-H6, Fig. 2.2) of different cross-shore extents. The shelf-indenting Carchuna Canyon and the undulating infralittoral prograding wedge (IPW) modulate the wave energy in the nearshore (see Chap. 3). The results obtained in Chap. 3 confirm that the concentration and divergence wave patterns are the main mechanisms responsible for the formation of the large-scale shoreline features [31]. In particular, the canyon is mainly responsible for the morphology in the western part of the beach (H1-H3), whereas the submerged IPW undulations modulate the alongshore wave energy and reinforce horns H4-H5-H6. Other mechanisms, such as the atmospheric-hydrodynamic coupling in the nearshore induced by Cape Sacratif [32], may reinforce this morphology.

Ortega-Sánchez et al. [31] reported the existence of rhythmic shoreline features spaced on the order of 10^2 m between H5 and H6. These researchers showed that one possible mechanism responsible for the formation of these features is the partial reflection of edge waves at the horns. Quevedo et al. [35] presented a hydrodynamic model that also predicts the generation of these beach lobes by turbulent wind vortices blowing over the sea surface on the lee side of geographic obstacles. As showed in Chap. 3, the IPW undulations between H5 and H6 induce an alongshore wave height modulation that may reinforce the formation of these beach lobes. The smaller cross- and alongshore dimensions of these beach lobes, compared with the dimensions of the horns H1-H6, may be due to the smaller propagation coefficients of both the WSW and E waves in this part of the beach.

The fact that the wave energy increases in the embayments of the large-scale features, regardless of the direction of the incoming predominant waves, in combination with the obliquity of the waves in the nearshore (Fig. 4.5), leads to the hypothesis that the submerged IPW undulations may play a key role in the formation of the shoreline morphology. This hypothesis is consistent with the results reported by [31], although they did not identify the bathymetric element causing the modulation. Whereas other studies have quantified the importance of the complexity of the inner shelf bathymetry in the nearshore hydrodynamics [2, 13], our results show that the adjacent bathymetry also has a strong influence on the evolution of the shoreline morphology.

Although the alongshore distribution of the wave energy may explain the formation (and location) and the irregular spacing of horns H2-H6, the main processes responsible for their different shapes remain unknown. As a first approximation, [34] applied the model developed by [3, 4] and found good general agreement between the large-scale features of Carchuna Beach and the results obtained using the [3] model. According to this model, instabilities and rhythmic features tend to appear on coastlines where the waves impinge with an obliquity greater than 45° with respect to the shore-normal orientation. Nevertheless, there were some aspects that were not reproduced by the model: (1) the protruding horns do not have a symmetric shape, (2) there is a general alongshore variation in the shape of the features, and (3) the relative orientation of the shoreline changes between features. Ortega-Sánchez et al. [34] concluded that nearshore wave energy modulation should be considered to fully explain the morphology of Carchuna Beach.

The complexity of the nearshore wave propagation patterns at Carchuna Beach, in combination with the curvature of the shoreline, results in an alongshore variation in the wave properties (i.e. Fig. 4.3). López-Ruiz et al. [24] developed a one-line-type

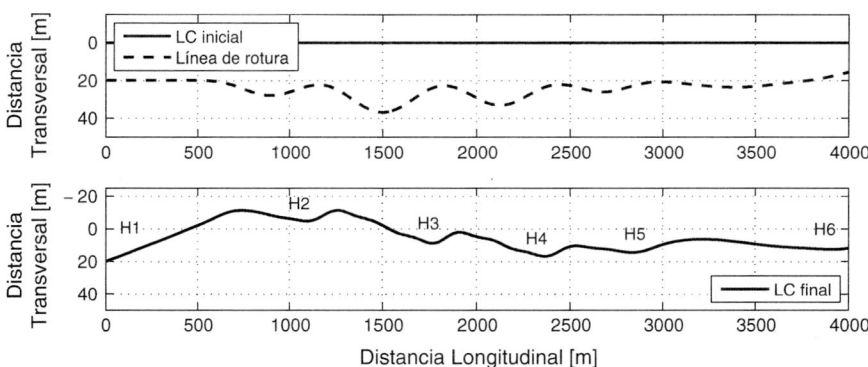

Fig. 4.7 *Upper panel*: Initial shoreline for the one-line simulation (*solid black line*) and surf zone width used during the simulation (*dashed blue line*). *Lower panel*: Shoreline obtained after the simulation. (*Source* [33]. Reproduced with permission of Elsevier)

model that considers the effects of the alongshore variation in the wave properties and the curvature of the shoreline. This model was applied to an initially rectilinear beach of the same length as Carchuna Beach. The wave forcing consisted of a modeled surf zone width that varied according to the wave propagation results described in Chap. 3, whereas the LST was obtained with the updated expression described throughout this chapter. Considering that the canyon is located in the western part of the beach and that westerly waves induce greater alongshore energy variations than easterly waves, only storm WSW waves were simulated.

Figure 4.7 shows the simulation results for the shoreline after 50 days of constant storm conditions. A series of horns with alongshore spacings very similar to those observed in Carchuna Beach were obtained. Moreover, the resulting features have an asymmetric shape: the side exposed to the more energetic westerly waves was larger than the protected side. The horns also have a non-uniform configuration in the alongshore direction, with changes in the relative orientation of the shoreline between the features due to the alongshore variation in wave energy.

Although the model captures the main geometric characteristics of the shoreline features, the shoreline evolution must be simulated using real variable climate forcing to reproduce the cross-shore length of the horns [24]. So, these qualitative results appear to confirm the hypothesis that the nearshore wave energy modulations induced by the presence of the submerged features under the prevailing waves can explain the development of the shoreline horns at Carchuna Beach. Moreover, these qualitative results appear to confirm that complex bathymetries exhibiting submerged large scale sedimentary features, i.e., IPW or shelf-indenting canyons, can play a key role in the formation of asymmetrical and/or irregular shoreline shapes, such as observed at Carchuna Beach. Simple shoreline evolution models (i.e., one-line type models) can be of significant use in the study of such shoreline shapes because they generally include both the wave energy patterns induced by the complex bathymetry and the changes in shoreline orientation as the coastline evolves.

4.3 Coastline Evolution Due to Delta Retreat: The Case of Playa Granada

The LST expression previously derived was also applied to Playa Granada. In this case, the expression was applied with two different types of wave conditions: (1) a synthetic dataset of conditions to describe the LST behavior along the delta for low energy and storm conditions and (2) the complete time series of wave conditions for the Epochs 1, 2 and 3, defined as the intervals between the dates in which bathymetric data were available (1999, 2004, 2008 and 2014, see Chaps. 2 and 3). The influence of wave directionality and both the volumetric changes and shoreline evolution of the delta were analyzed in detail with these time series.

4.3.1 Results for Low Energy and Storm Conditions

In the case of the synthetic dataset, the deep-water wave angle and the spectral peak period shown in Table 3.2 were used for both low energy and storm conditions. The shoreline angle and beach slope distributions were obtained from the bathymetric data of 1999, 2004, 2008 and 2014. The surf zone width was obtained from the results of the propagation model described in Chap. 3, obtaining the breaking point for 250 beach profiles equally distributed along the shoreline and defined as shore-normal.

Figure 4.8 depicts the results of LST obtained for low energy and storm conditions (Table 3.2). As a general spatial trend, the gradients in LST are important mainly at the Guadalfeo River mouth. It is also observed that greater values for LST are obtained under westerly waves, for both low energy and storm conditions. Regarding the low energy conditions, under westerly waves (Fig. 4.8c), the gradients in LST at the Guadalfeo River mouth are of minor importance for all bathymetries except for 2004, just after the establishment of the dam. In this case, gradients are more than 5 times greater than those for the other bathymetries. For easterly waves (Fig. 4.8c), the trend is very similar, with the only difference being that LST at the east side of the river mouth is more irregular than for westerly waves.

For storm conditions (Fig. 4.8d–f), LST patterns are different from those for low energy conditions. Significant differences are also observed between westerly and easterly wave conditions. Under westerly storms, the longshore gradients are significantly larger than those obtained for easterly storms, particularly for the 2008 and 2014 bathymetries. However, for easterly storms, the behavior is clearly different: although for 1999 the longshore gradients were important, they increased by a factor of two in 2004. These gradients vanished almost completely for 2008 and 2014, with the shoreline much more in equilibrium for such conditions. This indicates that the delta has acquired a shape that minimizes LST gradients for all wave conditions except for westerly storms, which are the most energetic.

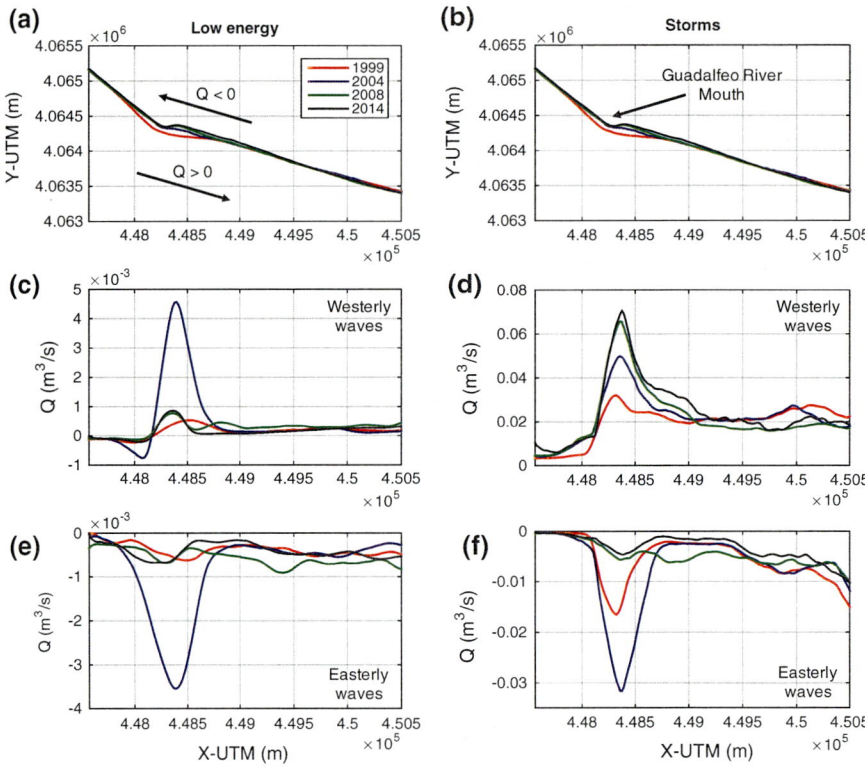

Fig. 4.8 Alongshore evolution of the longshore sediment transport: Low energy conditions for south-westerly **c** and easterly waves **e**; storm conditions for south-westerly waves **d** and easterly waves **f**. Shorelines of the four bathymetries are depicted in panels a and b. (*Source* [7]. Reproduced with permission of Elsevier)

4.3.2 LST and Wave Directionality

To analyze the role of wave directionality on LST, the complete time series of LST was obtained for the Epochs 1, 2 and 3. In order to improve the computational efficiency, downscaling techniques were applied: firstly, a database of representative wave conditions (H_0, T_p and θ) was generated using the downscaling method presented by [9, 10]. This first step synthesized the total dataset of deep water wave climate in a group of 280 sea states representing mild, mean and extreme wave conditions. These sea states ranged in the intervals $H_0 \in (0.1, 5.2)$ m, $T_p \in (2.5, 19)$ s, and $\theta \in (0, 360°)$ not equally distributed to account for the most likely sea states. Secondly, this database was propagated over the initial bathymetry for each epoch using the calibrated wave propagation model (see Chap. 3). With these propagations, nearshore breaking wave parameters are obtained and used to compute LST for every sea state of the database. Finally, these results allow reconstructing LST during the full study period by means of interpolation, obtaining LST along the coast for each

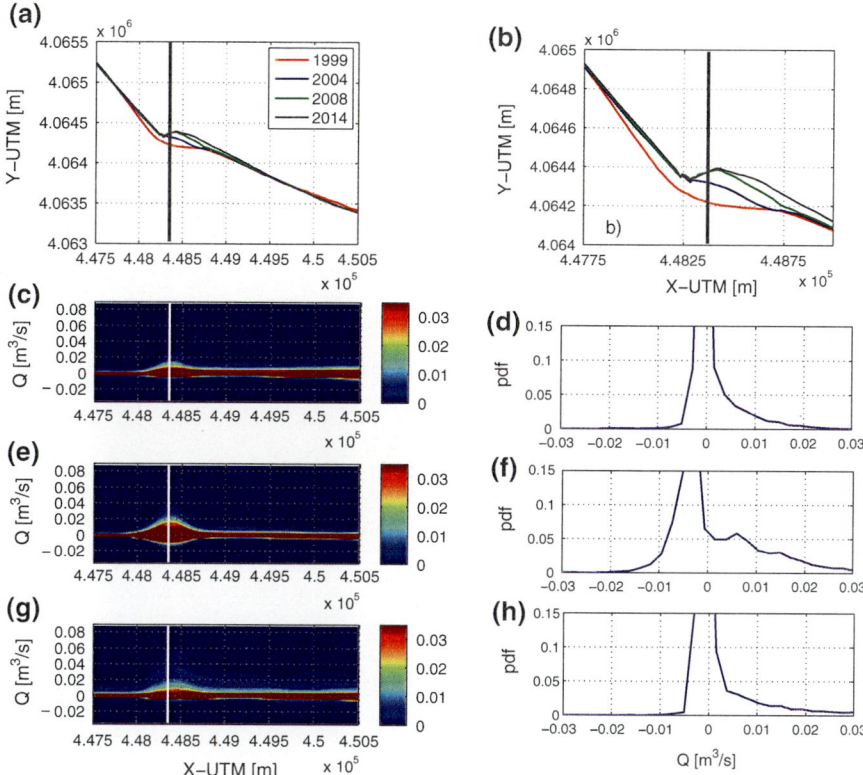

Fig. 4.9 Probability distribution functions (pdfs) for the LST during the Epochs 1 (**c**), 2 (**e**) and 3 (**g**). Colors correspond to the probability of values of LST (vertical axis) in its alongshore locations (horizontal axis). Panels d, f and h show the pdfs for a cross-shore section located at the vicinity of the river mouth. The location of this section is depicted, jointly with shoreline geometries, in panels a and b. (*Source* [7]. Reproduced with permission of Elsevier)

combination of offshore wave conditions. This reduces significantly the computational effort, and new LST results can be obtained without performing new wave propagations.

The results are shown in Fig. 4.9, where the probability distribution functions (pdfs) of LST are depicted for each epoch, depending on the location along the shoreline. The limits of the probability values were adjusted to appreciate the distribution of LST associated to low probabilities (pdf < 0.035). Moreover, the pdf for a specific cross-shore profile in the vicinity of the river mouth is also presented. This location corresponds to the beach profile in which the greatest values of LST are more likely. The aim of this analysis was to find out trends between the directionality distribution of waves and LST during each period.

As a general trend for the period 1999–2014, higher values of LST for both drift directions were clearly more likely in the vicinity of the river mouth. However, some differences in the pdfs are observed between the three epochs. During Epoch 1 (Fig. 4.9c–d), LST values with probabilities over 0.035 (red colors) barely present significant longshore gradients. On the contrary, LST with probabilities around 0.01 presented significant longshore gradients and values of $Q \simeq 0.02$ m³/s for westerly drifts close to the river mouth. Hence, LST related with westerly waves ($Q > 0$, westerly drift) was more frequent and intense than for easterly waves. These results coincided with those of wave directionality, as the value of the mean $P_{0,EW}$ for this epoch was over the mean for the entire dataset, indicating that this period was less easterly dominated than the average.

For Epoch 2 (Fig. 4.9e, f), higher values of LST and the associated longshore gradients in the vicinity of the river mouth were clearly more frequent than for Epoch 1. Although during this epoch LST with absolute values over 0.02 m³/s were more likely for westerly drifts, implying that significant longshore gradients in LST were more frequent for westerly waves, the symmetry of the deep red colors in the river mouth indicates that the majority of the LST values in the area of the river mouth were concentrated in $Q \in (-0.015, 0.015)$ m³/s. This implies that easterly waves were more important in this epoch, from a morphodynamic point of view, than in previous one. This is in clearly agreement with the results obtained for the yearly averaged $P_{0,EW}$ (Fig. 4.10c–e), as this period was the one with the lowest mean $P_{0,EW}$ value. For Epoch 3, the LST distribution is again very asymmetric with an important dominance of westerly drifts. However, in this case the highest LST rates ($Q > 0.03$ m³/s) were much more likely than in the previous epochs, with pdf > 0.01.

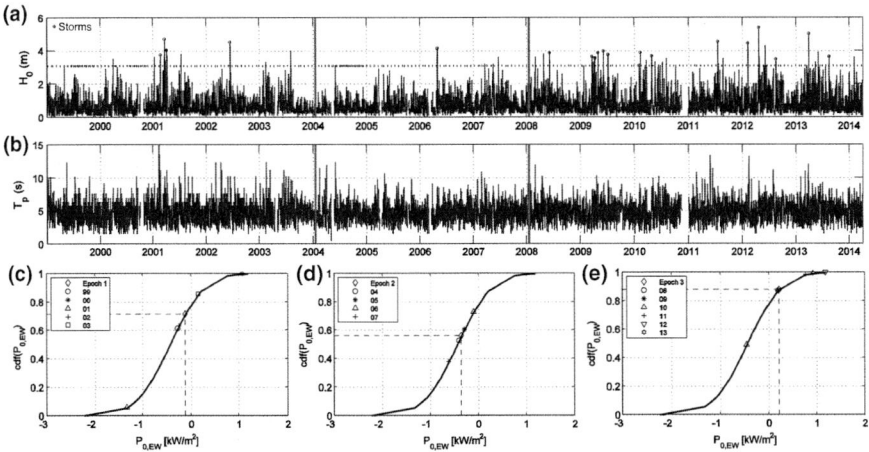

Fig. 4.10 Evolution of the deep-water wave height **a** and spectral peak period **b** from 1999 to 2014. Fitted cumulative distribution function (**cdf**) of the yearly averaged $P_{0,EW}$ (Normal distribution, $\mu = -0.45$ kW/m², $\sigma = 0.56$ kW/m²). *Black* markers correspond to values of the years in the Epoch 1 **c**, Epoch 2 **d** and Epoch 3 **e**. *Red* markers indicate the mean value for each epoch. (*Source* [7]. Reproduced with permission of Elsevier)

Fig. 4.11 **a** Shoreline at the beginning of each Epoch. **b** Averaged LST rates for Epochs 1 (*red*), 1 (*blue*) and 2 (*green*). **c** Alongshore evolution of LST volumetric changes in the nearshore during Epochs 1 (*red*), 1 (*blue*) and 2 (*green*). (*Source* [7]. Reproduced with permission of Elsevier)

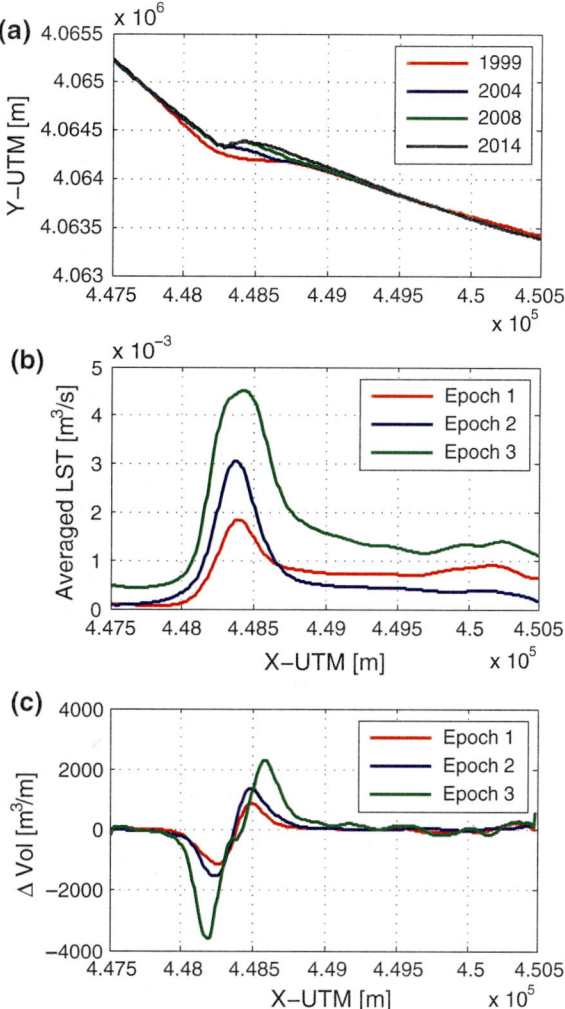

4.3.3 Volumetric Changes Induced by LST—Comparison with Delta Evolution

Figure 4.11 shows the observed coastline changes, the average LST rates and the volumetric differences based on the gradient in the LST for the three epochs. Since the LST formulation accounts for sediment transport in the breaking zone, the changes in sediment volume refer to changes in the shallower nearshore zone. The LST volume gradients, which are based on the potential littoral drift rates (Fig. 4.11b), predict erosion of the western flank of the delta and accretion of the eastern flank (Fig. 4.11c). However, higher eroded volumes were measured on the eastern flank

Fig. 4.12 Aerial images of the Guadalfeo River mouth taken in 1999 **a**, 2002 **b**, 2004 **c**, 2006 **d**, 2007 **e**, 2008 **f**, 2010 **g** and 2013 **h**. The identified shorelines are highlighted in colors. (*Source* [7]. Reproduced with permission of Elsevier)

Table 4.1 Sediment volumes (in m³) during Epochs 1, 2 and 3: total mobilized by LST, balance of LST, and measured between the shoreline and the maximum breaking depth (h_b)

	Epoch 1	Epoch 2	Epoch 3
LST mobilized	480,870	640,790	1,315,000
LST balance	$-89,997$	$-12,016$	$-119,623$
Measured $(0 - h_b)$	$-71,461$	$-18,198$	$-175,310$

during the three epochs (Fig. 3.13) and this is explained because LST volumes patterns were not realized because of the trapping of eastward moving sediment by the river jetties (when they extend into the sea). This occurred over most of the study period (Fig. 4.12), thereby reducing erosion on the western flank and causing erosion rather than accretion on the eastern flank. This highlights the importance of the river jetties in reversing the morphodynamic behavior expected for an unmodified beach.

The total sediment volume losses along the stretch of beach between Salobreña Rock and *Punta del Santo* (Fig. 3.5) induced by LST were approximately 90,000; 12,000 and 120,000 m³ for Epochs 1, 2 and 3, respectively (Table 4.1). These modelled values are similar (differences lower than 35%) than the measured volumetric changes between the shoreline and the maximum breaking depth for each epoch ($\sim -70,000$; $-18,000$ and $-175,000$ m³, respectively) and indicate that LST is the main contributor to coastal changes. The uncertainty associated with both measurements and modelling, particularly the lack of available multibeam bathymetries at a larger temporal resolution, should be considered to understand the obtained differences.

The negative sediment balance in the studied stretch of beach contrasts with the accumulation of sediments and the advance seaward of the coastline in the section *Punta del Santo*—Motril Port (Fig. 3.5), detailed in [6]. This is mainly due to the greater LST rates under westerly waves (Fig. 4.8) and the resulting dominance of westerly drift (Fig. 4.11b). The total mobilized sediment volumes modelled for each epoch ($V_3 > V_2 > V_1$, Table 4.1) are also consistent with both the wave directionality analysis performed (Fig. 4.10c–e) and the nearshore volumetric changes observed (Fig. 3.13).

4.4   Conclusions

1. A new, simple approach for calculating the LST on curvilinear coasts such as those of Carchuna Beach and Playa Granada was presented. Although previous approaches considered the shoreline curvature and variations in wave conditions [3], our approach can also consider variations in beach slope, shoreline geometry and sediment size. Numerical and observational results indicate that these variations are of major importance on mixed sand and gravel coasts.

2. We applied a one-line-type model based on the presented LST expression for Carchuna Beach. This model included the effects of the alongshore variation in wave properties and the curvature of the shoreline on an initial rectilinear beach with the same length as Carchuna. The model reproduced a series of horns with alongshore spacings that were very similar to those observed in Carchuna. Moreover, the model also reproduced the asymmetric shape of the features and the non-uniform alongshore configuration. Changes in the relative orientation of the shoreline between the features were attributed to alongshore variations in wave energy. Our results appear to confirm that complex bathymetries that exhibit submerged large-scale sedimentary features or shelf-indenting canyons may play a key role in the formation of asymmetrical and/or irregular shoreline shapes, such as were observed at Carchuna Beach.
3. In the case of Playa Granada, the LST on the coast of the delta was significantly modified by the combined effect of changes in the nearshore bathymetry (partly induced by river damming) and wave conditions (mainly wave directionality). Volumetric changes obtained using the LST gradients could quantitatively explain most of the measured volumetric changes (differences lower than 35%), which indicates that LST was the main driver of the nearshore changes in the delta.

Acknowledgements This work was partially supported by the Spanish Ministry of Economy and Competitiveness (Projects CTM2012-32439 and BIA2015-65598-P). The second author was funded by the Spanish Ministry of Economy and Competitiveness (Research Contract BES-2013-062617 and Mobility Grant EEBB-I-16-11009). The authors are indebted to Gerd Masselink for his valuable suggestions and comments. Prof. Baquerizo is also acknowledged for helping in the development the LST expression for curvilinear coasts. We also thank Fátima Pereira for her support creating the bathymetric scenarios of Carchuna Beach and Cristóbal Rodríguez-Delgado for his assistance with the bathymetric data processing of Playa Granada.

References

1. Aarninkhof, S.G.J., Holman, R.A.: Monitoring the nearshore with video. Backscatter **10**, 8–11 (1999)
2. Apotsos, A., Raubenheimer, B., Elgar, S., Guza, R.T.: Wave-driven setup and along- shore flows observed onshore of a submarine canyon. J. Geophys. Res. **113**, C07,025 (2008)
3. Ashton, A., Murray, A.B.: High-angle wave instability and emergent shoreline shapes: 1 Modeling of sand waves, flying spits, and capes. J. Geophys. Res. **111**, F04,011 (2006). doi:10.1029/2005JF000423
4. Ashton, A., Murray, A.B.: High-angle wave instability and emergent shoreline shapes: 2 Wave climate analysis and comparisons to nature. J. Geophys. Res. **111**, F04,012 (2006). doi:10.1029/2005JF000422
5. Bayram, A., Larson, M., Hanson, H.: A new formula for the total longshore sediment transport rate. Coast. Eng. **540**(9), 700–710 (2007)
6. Bergillos, R.J., Delgado-Rodríguez, C., López-Ruiz, A., Millares, A., Ortega-Sánchez, M., Losada, M.A.: Recent human-induced coastal changes in the Guadalfeo river deltaic system (southern Spain). In: Proceedings of the 36th IAHR-International Association for Hydro-Environment Engineering and Research World Congress (2015)

7. Bergillos, R.J., López-Ruiz, A., Ortega-Sánchez, M., Masselink, G., Losada, M.A.: Implications of delta retreat on wave propagation and longshore sediment transport—Guadalfeo case study (southern Spain). Mar. Geol. **382**, 1–16 (2016)
8. Booij, N., Ris, R.C., Holthuijsen, L.H.: A third-generation wave model for coastal regions: 1. Model description and validation. J. Geophys. Res.: Oceans **104**(C4), 7649–7666 (1999)
9. Camus, P., Mendez, F.J., Medina, R.: A hybrid efficient method to downscale wave climate to coastal areas. Coast. Eng. **58**(9), 851–862 (2011)
10. Camus, P., Menéndez, M., Méndez, F.J., Izaguirre, C., Espejo, A., Cánovas, V., Pérez, J., Rueda, A., Losada, I.J., Medina, R.: A weather-type statistical downscaling framework for ocean wave climate. J. Geophys. Res. C: Oceans **119**(11), 7389–7405 (2014)
11. Dean, R.G., Dalrymple, R.A.: Coastal Processes with Engineering Applications. Cambridge University Press (2002)
12. Fredsøe, J., Deigaard, R.: Mechanics of Coastal Sediment Transport. World Scientific (1992)
13. Gorrell, L., Raubenheimer, B., Elgar, S., Guza, R.T.: SWAN predictions of waves observed in shallow water onshore of complex bathymetry. Coast. Eng. **58**, 510–516 (2011)
14. Hanson, H.: GENESIS—A generalized shoreline change numerical model. J. Coast. Res. **50**(1), 1–27 (1989)
15. Harley, M.D., Turner, I.L., Short, A.D., Ranasinghe, R.: Assessment and integration of conventional, RTK-GPS and image-derived beach survey methods for daily to decadal coastal monitoring. Coast. Eng. **580**(2), 194–205 (2011)
16. Holland, K.T., Holman, R.A., Lippmann, T.C., Stanley, J., Plant, N.: Practical use of video imagery in nearshore oceanographic field studies. IEEE J. Oceanic Eng. **22**(1), 81–92 (1997)
17. Hydraulics, W.: UNIBEST, a software suite for simulation of sediment transport processes and related morphodynamics of beach profiles and coastline evolution. Model description and validation. Technical report, Delft hydraulics Report H454.14 (1992)
18. Inman, D.L., Bagnold, R.A.: Littoral Processes. In: The Sea, Vol. 3. Wiley, New York (1963)
19. Kamphuis, J.W.: Alongshore sediment transport rate. J. Waterw Port Coast. Ocean Eng. **117**(6), 624 (1991)
20. Kamphuis, J.W.: Alongshore transport of sand. In: Proceedings of the 28th International Conference on Coastal Engineering, pp. 2330–2345. ASCE (2002)
21. Komar, P.D.: Beach Processes and Sedimentation, 2nd edn. Prentice-Hall, Upper Saddle River, NJ (1998)
22. Larson, M., Kraus, N.C.: Prediction of cross-shore sediment transport at different spatial and temporal scales. Mar. Geol. **126**(1–4), 111–127 (1995)
23. Longuet-Higgins, M.S.: Longshore currents generated by obliquely incident sea waves, 1. J. Geophys. Res. **75**(33), 6778–6789 (1970)
24. López-Ruiz, A., Ortega-Sánchez, M., Baquerizo, A., Losada, M.A.: Short and medium- term evolution of shoreline undulations on curvilinear coasts. Geomorphology **159**, 189–200 (2012)
25. López-Ruiz, A., Ortega-Sánchez, M., Baquerizo, A., Losada, M.A.: A note on along- shore sediment transport on weakly curvilinear coasts and its implications. Coast. Eng. **88**, 143–153 (2014)
26. López-Ruiz, A., Ortega-Sánchez, M., Baquerizo, A., Navidad, D., Losada, M.A.: Nonuniform alongshore sediment transport induced by coastline curvature. In: 33th Coastal Engineering Conference, Electronic edition, pp. 3046–3052. Texas Digital Library (2012)
27. López-Ruiz, A., Solari, S., Ortega-Sánchez, M., Losada, M.A.: A simple approximation for wave refraction—Application to the assessment of the nearshore wave directionality. Ocean Modell. **96**, 324–333 (2015)
28. McLaren, P., Bowles, D.: The effects of sediment transport on grain-size distributions. J. Sediment. Petrol. **55**, 457–470 (1985)
29. Mil-Homens, J., Ranasinghe, R., de Vries, J.V.T., Stive, M.J.F.: Re-evaluation and improvement of three commonly used bulk longshore sediment transport formulas. Coast. Eng. **75**, 29–39 (2013)
30. Noda, E.K.: Wave-induced nearshore circulation. J. Geophys. Res. **79**, 4097–4106 (1974)

31. Ortega-Sánchez, M., Baquerizo, A., Losada, M.A.: On the development of large-scale cuspate features on a semi-reflective beach: Carchuna beach, Southern Spain. Mar. Geol. **198**, 209–223 (2003)
32. Ortega-Sánchez, M., Bramato, S., Quevedo, E., Mans, C., Losada, M.A.: Atmospheric-hydrodynamic coupling in the nearshore. Geophys. Res. Lett. **35**, L23,601 (2008)
33. Ortega-Sánchez, M., Lobo, F.J., López-Ruiz, A., Losada, M.A., Fernández-Salas, L.M.: The influence of shelf-indenting canyons and infralittoral prograding wedges on coastal morphology: the Carchuna system in Southern Spain. Mar. Geol. **347**, 107–122 (2014)
34. Ortega-Sánchez, M., Quevedo, E., Baquerizo, A., Losada, M.A.: Comment on "High-angle wave instability and emergent shoreline shapes: 1 Modeling of sand waves, flying spits, and capes by Ashton, Andrew D., Brad Murray, A". J. Geophys. Res. **113**, F01,005 (2008)
35. Quevedo, E., Baquerizo, A., Losada, M.A., Ortega-Sánchez, M.: Large-scale coastal features generated by atmospheric pulses and associated edge waves. Mar. Geol. **247**, 226–236 (2008)
36. Ranasinghe, R., Pattiaratchi, C.: The seasonal closure of tidal inlets: Wilson inlet—a case study. Coast. Eng. **370**(1), 37–56 (1999)
37. Reeve, D., Chadwick, A., Fleming, C.: Coastal Engineering: Processes, Theory and Design Practice (2nd edn.). Spon Press (2012)
38. Schoonees, J.S., Theron, A.K.: Review of the field-data base for longshore sediment transport. Coast. Eng. **190**(1–2), 1–25 (1993)
39. Short, A.D.: Handbook of Beach and Shoreface Morphodynamics. Wiley (2000)
40. U.S.A.C.E.: Shore Protection Manual. U.S. Government Printing Office, Washington, D.C. (1984)
41. del Valle, R., Medina, R., Losada, M.A.: Dependence of coefficient K on grain size. J. Waterw. Port Coast. Ocean Eng. **119**(5), 568–574 (1993)

Chapter 5
Morpho-Sedimentary Dynamics of Mixed Sand and Gravel Coasts

Abstract This chapter addresses the changes in the morphology and sedimentology of a micro-tidal mixed sand and gravel beach (Playa Granada, southern Spain)
forced by wave and water-level variations, and human intervention through nourishment. Monthly and storm event-driven beach surveys, consisting of topographical
measurements and sediment sampling in two selected areas, were carried out over
a one-year period. Three prevailing sediment fractions (sand, fine gravel and coarse
gravel) and two end-member morphological states of the upper beach profile (con-
vex with multiple berms and concave with a single storm berm) were identified.
Between them, several transitional profiles were formed, characterized by developing berms that progressively overlapped, generating sediment variability both across
the beach profile and with depth. The results indicate that the total run-up (including
water-level) reached during an event represents a more accurate threshold for differ-
entiating between erosional and depositional conditions than wave height. They also
suggest that mixed sand and gravel beaches recover faster from storm erosion than
sandy beaches. The long-term benefit of the artificial nourishment that took place at
the end of the survey period was very limited and this is attributed to the too fine
sediment used for the nourishment and its placement too high on the beach. Clearly,
nourishment interventions must take into account the natural sediment distribution
and the profile shape to avoid rapid losses of the nourished sediment.

5.1 Methodology

5.1.1 Data

5.1.1.1 Maritime Data

One year of hourly hindcasted data, corresponding to SIMAR point number 2041080,
was used for driving the wave-induced coastal morphological changes. The following
variables were extracted from these hindcasted data: deep-water significant wave
height (H_0), spectral peak period (T_p), deep-water wave direction (θ_0), wind velocity

© The Author(s) 2017   63
M. Ortega-Sánchez et al., *Morphodynamics of Mediterranean Mixed
Sand and Gravel Coasts*, SpringerBriefs in Earth Sciences,
DOI 10.1007/978-3-319-52440-5_5

(V_w) and wind direction (θ_w). Furthermore, the astronomical tide measured by a gauge located in the Motril Port was used to represent the tidal forcing. Both wave and tide data were provided by *Puertos del Estado*.

5.1.1.2   Field Surveys

Field surveys were performed from October 2013 to September 2014 (hereafter referred to as the study period). To analyze the beach morphological evolution, two study areas within the study site were selected (Fig. 5.1) to ensure the results were representative for entire beach section.

Monthly periodic field surveys were performed during the study period, consisting of topographic and sediment sizes measurements of the beach profile. In addition, several specific surveys were carried out before and after two significant storms (December 2013 and March 2014), and before, during and after the artificial replenishment of the beach performed in June 2014 (Table 5.1). Each survey was carried out under low tide conditions and the measurements were referenced to the mean low water spring level (MLWS) to avoid negative contributions of the astronomical tide to the total run-up. The two major storms that occurred during the study period were identified by means of the peaks over threshold (POT) method considering $H_T = 3.1\,\text{m}$ $(H_{99.9\%})$ and storm durations lasting longer than 6 h.

The topography was recorded with a highly accurate DGPS (Javad Maxor) with less than 2 cm of both horizontal and vertical instrument errors. Previously, the geodesic coordinates of the vertex *105582 Punta del Santo* were moved to the positions of the GPS-base in the study areas. Ten profiles were measured at each study area to obtain an alongshore-averaged profile to reduce the uncertainty associated with measurement errors and alongshore variability (Fig. 5.1). Sediment samples, both at the surface and at depth (0–30 cm), were taken at three points of each profile (Fig. 5.1) to capture the spatial variability in the sediment distribution. Sieve analyses of the

Fig. 5.1 Selected study areas and geodesic vertex *Punta del Santo*. Ten profiles were measured in both study areas (*dashed lines*) and samples were taken both at the surface and at depth in each profile (dots). (*Source* [2]. Reproduced with permission of Elsevier)

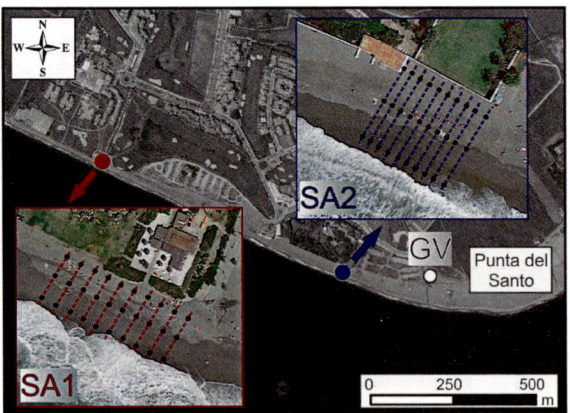

Table 5.1 Timeline of the periodic (P) and specific (S) surveys carried out during the study period

No. Survey	1	2	3	4	5	6	7	8	9
Date	25/10	22/11	20/12	27/12	10/1	21/1	27/2	10/3	17/3
Type of Survey	P	P	P	S	S	P	P	S	S
10	11	12	13	14	15	16	17	18	19
31/3	25/4	19/5	2/6	13/6	17/6	25/6	24/7	21/8	19/9
P	P	P	S	S	S	P	P	P	P

sampled sediments in each study area were performed following the basic methods of [6] with grain size nomenclature according to [22].

5.1.1.3 Bathymetry

A high-resolution multibeam bathymetric survey was carried out in October 2013 (beginning of the study period) at the study site. The data were acquired using DGPS navigation referring to the WGS-84 ellipsoid. Accurate navigation and real-time pitch, roll, and heave were corrected. The multibeam data were also corrected for the water column velocity. These bathymetric data were used as the bottom boundary condition for the wave propagation model (Sect. 5.1.2).

5.1.2 Wave Propagation Model

The WAVE module of the Delft3D model [10, 11], which is based on the spectral wave model SWAN [7], was applied (considering the SIMAR point data) to estimate inshore wave conditions. Simulated wave heights were obtained at points with depth of 8 m (H_{8m}) to avoid the influence of wave breaking and these inshore wave conditions were related to the beach response. The model domain consisted of two different grids, shown in Fig. 3.5. The first grid is a coarse curvilinear 82 × 82—cell grid covering the entire Playa Granada region, with cell sizes that decrease with depth from 88 × 60 to 48 × 35 m. The second grid is a nested grid with 82 and 144 cells in the alongshore and cross-shore directions, respectively, and cell sizes of about 25 × 14 m. This model was calibrated and successfully validated for the study site through comparison with field data by [1].

5.1.3 Total Run-Up and Sediment Mobility Formulations

5.1.3.1 Total Run-Up

The run-up measured on the beach by means of the DGPS (based on observations of run-up mark) was compared with estimates of the total run-up, obtained as the sum of astronomical tide, storm surge (wind set-up and inverse barometric effect) and wave run-up. The wind set-up was calculated as follows: $\Delta\eta_{wind} = \tau_{wind}/(\rho g h_0)\Delta x$ [3], where g is the acceleration of gravity, $\rho = 1025$ kg/m^3 is the density of salt water, the depth of the wave base level is represented by $h_0 = L_0/4$, ΔX is the wave fetch from the center of the low-pressure system to the coast (estimated through isobar maps) and the tangential wind stress is obtained from $\tau_{wind} = \rho_a U_*^2$, where ρ_a is the air density and U_* is the friction velocity. The barometric set-up was obtained from $\Delta\eta_{bar} = \Delta P_a/(\rho g)$ [5], where ΔP_a represents the atmospheric pressure variation relative to the long-term average pressure at Motril Port. Finally, the wave run-up was calculated through the equation $\Delta\eta_{wave} = 0.36\, g^{0.5}\, H_{8,0}^{0.5}\, T_p\, \tan\beta$ [17], where $\tan\beta$ is the intertidal slope and $H_{8,0}$ is the modeled wave height at 8 m water depth (H_{8m}) de-shoaled to deep water using linear theory and assuming parallel bottom contours. This parameter allows accounting for the alongshore variability of the inshore wave height and is consistent with the run-up expression which requires deep-water wave height. This formulation for total run-up was successfully used and compared with high resolution images from a video camera by [4] deployed on Carchuna Beach.

5.1.3.2 Sediment Mobility

[20] derived the following relationship to determine the accretion/erosion states of a beach:

$$\frac{H_{8,0}}{L_0} = C \left(\tan\bar{\beta}\right)^{-0.27} \left(\frac{D}{L_0}\right)^{0.67} \tag{5.1}$$

where $H_{8,0}/L_0$ is the deep-water wave steepness, L_0 is the deep-water wave length, $C = 18$ is an empirical constant, D is the grain size and $\tan\bar{\beta}$ is the average nearshore bottom slope to a water depth of 20 m. According to [20], the beach erodes (accretes) when the left-hand side of the equation is greater (smaller) than the right-hand side. The difference (S_r) between both sides of the equation was calculated considering the prevailing sediment sizes measured in Playa Granada for each sea state and the results were related with the beach evolution. This equation, deduced for sandy beaches, was tested in the study site to analyze the role of the different fractions and to discuss its applicability to MSGBs.

5.2 Wave, Wind and Water-Level Conditions—Contributions to the Total Run-Up

Figure 5.2 depicts the evolution of the wave and wind conditions during the study period. The deep-water significant wave height and spectral peak period were generally $H_0 < 1$ m (73% of the time) and $T_p < 6$ s (76% of the time), indicating that the beach predominantly experienced low energy waves. This agrees with the generally low energy wave climate of this part of the Mediterranean Spanish coast [18]. The predominant deep-water wave directions were west-southwest and east-southeast, and the maximum T_p was 12 s and associated with easterly waves (Fig. 5.2). This relatively high value for T_p (for the Mediterranean) has been exceeded 0.24% of the time since 1958. The prevailing wind velocity was less than 10 m/s with incoming directions from the east-southeast and west-southwest. The latter was more frequent and was generally associated with higher velocities. The wind direction was closely related to the wave direction.

Applying the POT method, two storms occurred that had maximum H_0 of 4.6 and 4 m, and maximum H_{8m} of 3.6 and 3.2 m in study area 1 and H_{8m} of 3.8 and 3.3 m in study area 2. Both storms were associated with westerly waves ($\theta_0 \sim 240°$, with maximum T_p of 9.6 s and 8.8 s (Fig. 5.2). The maximum V_w during storms 1 and 2 was 19.4 m/s and the θ_w was $\sim 260°$. Both two storm events had a very high energy content compared to other storms that occurred in the Alborán Sea. Specifically,

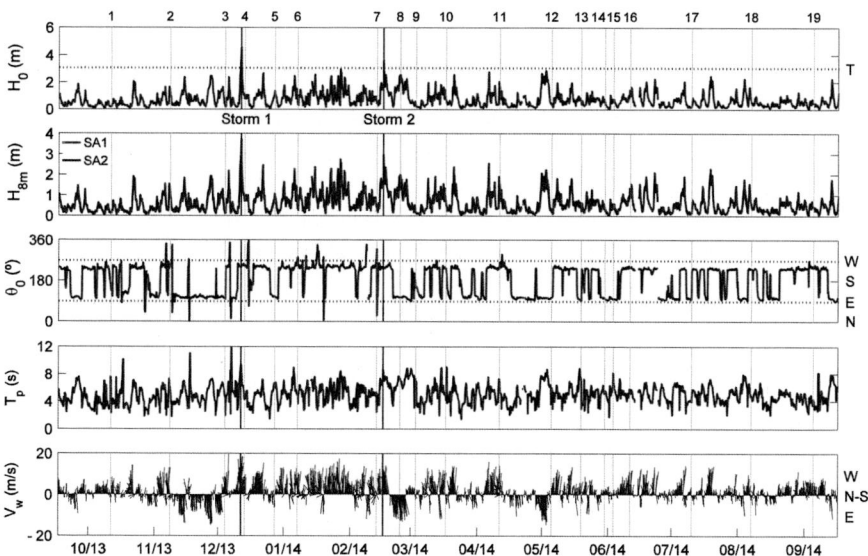

Fig. 5.2 From *top* to *bottom*: evolution of the deep-water wave height, wave height at depths of 8 m (*red*: study area 1, *blue*: study area 2), deep-water wave direction, spectral peak period, and wind velocity and direction. The vertical lines (*gray*) indicate the date of the field surveys and storms are marked in brown. (*Source* [2]. Reproduced with permission of Elsevier)

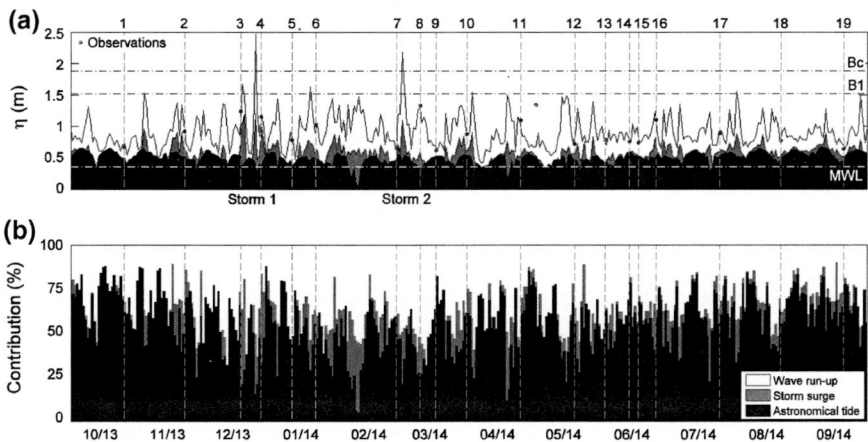

Fig. 5.3 **a** Evolution of the astronomical tide, storm surge and wave run-up and **b** contribution of each in the study area 1 during the study period. $\eta = 0$ indicates the MLWS level and the vertical lines (*gray*) indicate the date of the field surveys. The observations of total run-up (*red circles*), height of the mean water-level (MWL), the beach crest (Bc) and the upper berm (B1) are indicated. (*Source* [2]. Reproduced with permission of Elsevier)

storm 1 was the third most severe since 1958 to the end of the study period, based on the SIMAR 2041080 data.

Figure 5.3a shows the time series of the maximum daily total run-up (referenced to the MLWS) in study area 1 estimated using the formulations described in Sect. 5.1.3.1; the relative contributions of the different contributors (astronomical tide, storm surge and wave run-up) to the total run-up are depicted in Fig. 5.3b. The results for study area 2 were very similar (differences less than 5%). Comparison of the estimated maximum run-up values with those observed during 21 field measurements (19 during field surveys and 2 during storms) yielded high correlation (differences less than 9%), inspiring confidence in the estimated total run-up time series.

The measured tidal ranges during storms 1 and 2 were 0.2 m and 0.44 m, respectively, and were considerably less than the maximum tide range of 0.6 m (Fig. 5.3a). This reinforces that the contribution of the astronomical tide to the total run-up under high-energy conditions is relatively minor, representing less than 21% for both storms (Fig. 5.3b). In addition to wave run-up, storm surge is also a significant contributor to the total run-up, contributing more than 30% during both storms (Fig. 5.3b). The contribution of wave run-up reached values of almost 55 and 70% after storms 1 and 2 (recovery phases), i.e., between surveys 4–5 and 8–9, respectively (Fig. 5.3b).

Waves are frequently considered as the main driver of changes in the profile of micro-tidal beaches. However, Fig. 5.3b indicates that storm surge resulting from low atmospheric pressure and wind stress can also be important contributors to the total elevation under storm conditions and, consequently, to the erosion of the beach. If wind velocities are high enough ($V_w \sim 15$ m/s) and pressure gradients are negative, the resulting large storm surge enables waves to reach the upper parts of the beach profile (backshore), as shown in Fig. 5.3a.

Table 5.2 Morphological characteristics of the natural and replenished profiles in the study areas. LE: low energy conditions, S: storms

		Study area 1	Study area 2
Natural profiles	Slope	0.05 (LE)–0.069 (S)	0.056 (LE)–0.073 (S)
	Beach width (m)	24.74 (S)–35 (LE)	39.5 (S)–50 (LE)
	Unit volume (m^2)	27 (S)–41.40 (LE)	51.19 (S)–73.56 (LE)
Replenished profiles	Slope	0.057–0.059	0.05–0.053
	Beach width (m)	33.22–34.72	47–50
	Unit volume (m^2)	39.17–45.38	77.88–87.69

5.3 Morphological Response of the Beach Profile—Role of Total Run-Up

A total of 190 upper profiles (beach profile above the MLWS level) were measured during the study period in each study area: 130 before the artificial replenishment (natural profile) and 60 both during the nourishment and afterwards (replenished). Table 5.2 shows that the beach width (cross-shore distance between the MLWS level and the nearest building) and unit volume (calculated by the trapezoidal rule) of the beach typically increases under low energy conditions (LE) and decreases after storms (S). The slope of the natural profile, defined by the ratio between the height of beach crest and the beach width, was $0.05 - 0.069$ and $0.056 - 0.073$ in study areas 1 and 2, respectively.

Figure 5.4 depicts both the maximum wave height and total run-up (including astronomical tide, storm surge and wave run-up) before each survey along with the sediment volume of the upper profile in both study areas (in m^3 per unit m beach length per day, or m^2/day). It is observed that beach erosion/accretion not only depends on wave height, but on the sum of the three components that contribute to the total run-up. Actually, a relationship between the maximum total run-up between surveys and the beach response is clearly observed, especially after the two storms (surveys 4 and 8).

The differences between the bed elevation in each survey and the average profile in study area 1 are also shown in Fig. 5.4 (lower panel). During storms 1 and 2, the erosion rates in study area 1 were 2.06 and 1.09 m^2/day, respectively; whereas they were 3.2 and 1.76 m^2/day in study area 2. If a beach overwashes (Fig. 5.5), erosion tends to be less, because the wave energy is dissipated across the backshore and sediment is retained within the beach in the form of overwash deposit [15, 16]. This occurred in both study areas during storms, as the entire beach was overwashed. On the other hand, the recovery rates after the storm 1 and 2 were at least 0.79 and 0.51 m^2/day in study area 1, and 1.76 and 0.94 in study area 2, respectively. It is important to highlight that these values are average rates between surveys, so the maximum erosion/accretion rates were most likely higher.

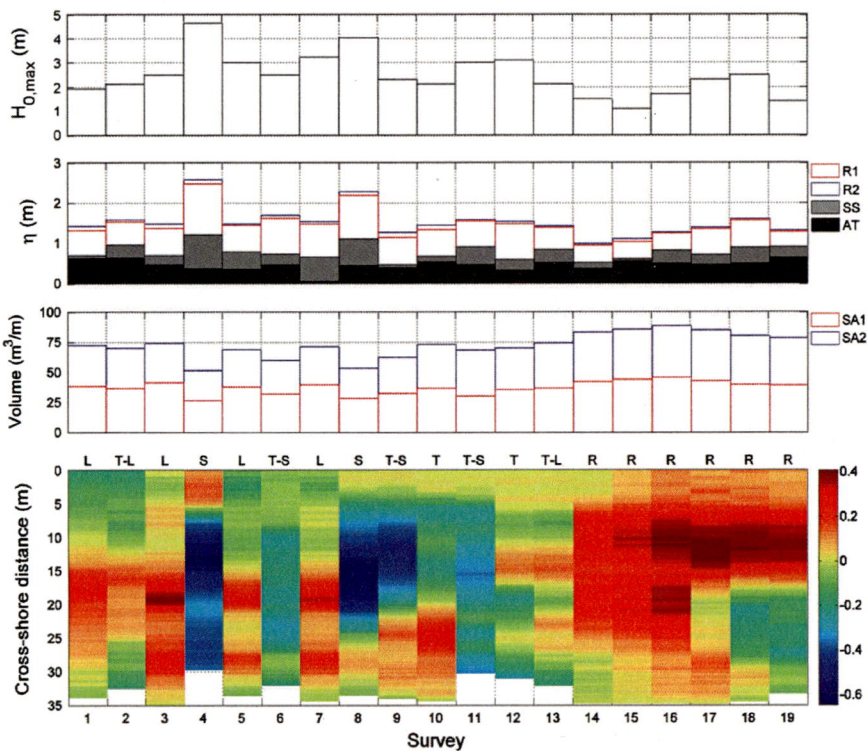

Fig. 5.4 From *top* to *bottom*: maximum deep-water wave height before each survey; astronomical tide (*black*), storm surge (*gray*) and wave run-up (*red*: study area 1, *blue*: study area 2) contributions to the maximum total run-up before each survey; unit volume of the upper profile (*red*: study area 1, *blue*: study area 2); and differences between the profile in each survey and the average profile in study area 1. States of the profile, according to Fig. 5.11, are shown. (*Source* [2]. Reproduced with permission of Elsevier)

The recovery rates (periods) in the study are after the third most severe storm since 1958 were significant higher (lower) than those detailed by [9, 19, 21], who measured average recovery rates of about 0.09, 0.11 and 0.26 m^2/day on the sandy beaches of Duck (US), Moruya (Australia) and Perranporth (UK), respectively. This supports the conclusions of [8, 13]: MSGBs may experience more active sediment transport than sandy beaches.

The profiles of both study areas were flattened due to the artificial nourishment carried out in June 2014. It consisted of an input of 8.4 and 14 m^3/m in study areas 1 and 2 over beach lengths equal to 500 and 300 m, respectively. The slope of the replenished profiles was slightly milder than those of the pre-nourished beach, but higher than the slope under low energy conditions (Table 5.2). The sediment volume of the replenished profiles was greater than that of most natural profiles, although the width was similar to those of the natural profiles under low energy conditions (Table 5.2).

Fig. 5.5 Photographs showing how the total run-up overwashes the beach profile. (*Source* Photographs by Miguel Ortega-Sánchez and Rafael J. Bergillos)

Figure 5.6 shows the evolution of the upper profile in study area 1 since the artificial replenishment to the end of the study period. Only one month after the artificial replenishment, and under prevailing low energy conditions and total run-up lower than 1.4 m until survey 17 (Figs. 5.2 and 5.3a), the unit volume loss was about 2.6 m^2 in study area 1 and 3.3 m^2 in study area 2. Berms started to appear due to the total run-up attained during this period (Figs. 5.4 and 5.6).

Between surveys 17 and 18, the profile shape also changed significantly in both study areas, but the variation was less between surveys 18 and 19 (Figs. 5.4 and 5.6), most likely due to the smaller magnitude of the forcing agents and the total run-up (Figs. 5.2 and 5.3a). The attenuation of the system response after the discharge of sediments could be another cause of this lower variation. However, not only wave processes, but also the gusts of wind after the nourishment project could contribute to the rapid loss of fine sediments, considering that wind velocities reached maximum values of 14.5 m/s and 13.5 m/s before surveys 17 and 18, respectively (Fig. 5.2).

Altogether, the beach width in study areas 1 and 2 decreased by approximately 4 and 6% in the months after the nourishment, respectively, whereas the unit volume loss was 6.2 m^3/m in study area 1 and 9.8 m^3/m in study area 2. Furthermore, despite the artificial replenishment, the unit volumes measured in study areas 1 and 2 in September 2014 (39.16 and 77.8 m^2) were similar to those measured in October 2013 (38.74 and 73.03 m^2), as shown in Fig. 5.4. Thus, the long-term benefit of the nourishment was very limited.

Fig. 5.6 Pre-nourished upper profile and evolution since the artificial replenishment until the end of the study period in study area 1. Height = 0 indicates the MLWS level. (*Source* [2]. Reproduced with permission of Elsevier)

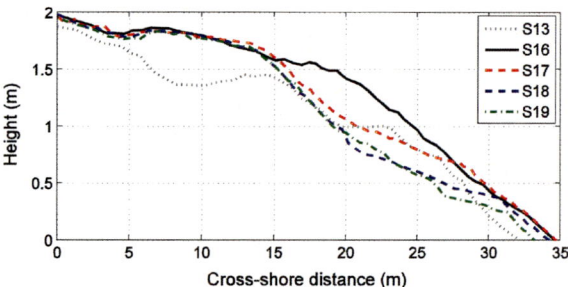

5.4 Beach Sediments

The average grain size distribution (based on all sediment samples) before the artificial replenishment (Fig. 5.7a) shows that three sediment fractions are predominant in Playa Granada: sand ($D_1 = 0.25 - 0.5$ mm, Fig. 5.7c), fine gravel ($D_2 = 2 - 8$ mm, Fig. 5.7d) and coarse gravel ($D_3 = 8 - 32$ mm, Fig. 5.7e). The foreshore (from the MLWS to the maximum total run-up reached under low energy conditions) showed greater sediment size variability than the backshore in both study areas, as shown in Sect. 5.5. In addition to this cross-shore variability, different levels of gradation at depth were also found. The sand-gravel ratio limits were $30 - 70$ and $36 - 64\%$ in study area 1, and $33 - 67$ and $23 - 77\%$ in study area 2.

The nourished material (Fig. 5.8a), shown in Fig. 5.7b, was significantly finer than the natural sediment ($D_{50} = 1.92$ mm *vs* $D_{50} = 4.35$ mm). Coarse sand ($1 - 2$ mm) and fine gravel ($2 - 8$ mm) dominated (Fig. 5.7f), with a sand-gravel ratio of about $52.5 - 47.5\%$. Thus, the sand fraction of the nourished sediment was higher than that of the native sediment. After the nourishment, the sand-gravel ratio progressively reduced from its initial value ($52.5 - 47.5\%$) to about $46.15 - 53.35\%$ and $41.65 - 58.35\%$ in study areas 1 and 2, respectively. Therefore, the reduction in the percentage of sand was higher in study area 2, where the unit volume loss was also higher (Table 5.2).

Figure 5.9 shows the results of applying the formulation of [20] during the study period. Considering the three prevailing sediment fractions in the study site (D_1, D_2 and D_3), the erosion ($S_r > 0$) and accretion ($S_r < 0$) states alternated for the sand fraction (Fig. 5.9, upper panel), while for the two gravel fractions only deposition states occurred (Fig. 5.9, middle and lower panels). These results are similar to those obtained by [4] through application of this formulation and observations based on high-resolution images for Carchuna Beach: sand was transported offshore during storms and beach recovery was limited to low-energy sea states, whereas only onshore migration took place for the gravels.

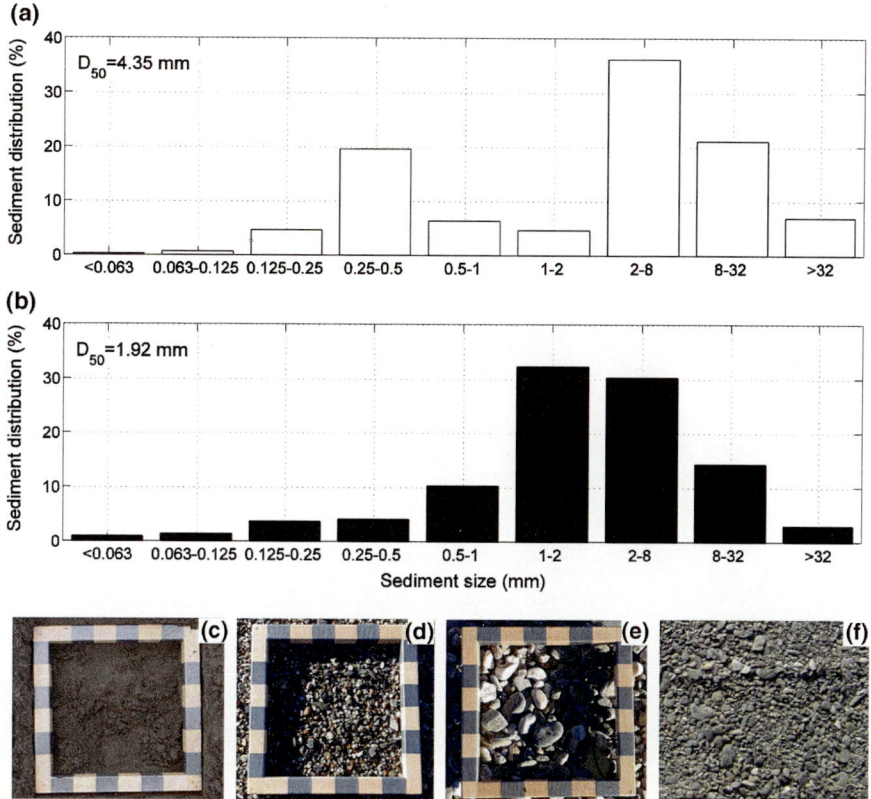

Fig. 5.7 Grain size histograms of the set of samples in both study areas before the artificial replenishment (**a**) and of the sediments used for the artificial replenishment (**b**); sand (**c**), fine gravel (**d**) and coarse gravel (**e**) natural fractions; and sediments supplied in June 2014 (**f**). The D_{50} of the nourishment was lower than that of the natural sediment. (*Source* [2]. Reproduced with permission of Elsevier)

5.5 Morpho-Sedimentary States of the Beach Profile

5.5.1 Low Energy State

Under prevailing low energy conditions, the upper profile in study area 1 has two berms (B1 and B2, Fig. 5.10a) composed of a surface layer of coarse gravels ($D_1 = 8 - 32$ mm), a subsurface layer of fine gravels ($D_2 = 2 - 8$ mm), and a layer of sand ($D_3 = 0.25 - 0.5$ mm) at the base of the deposit (Fig. 5.10a). This pattern is repeated at depth, probably reflecting previous berm deposits. The average percentage of sand-gravel along the sampled sediment layer is $35.8 - 64.2\%$. The backshore (cross-shore distance <15 m, Fig. 5.10a) consists mainly of sand (Table 5.3), whereas the composition of the sediment in the active swash zone is highly variable (sand-fine

(a)   **(b)**

Fig. 5.8 a Artificial replenishment done in June 2014, consisting of an input of sediment with uniform distribution. **b** Upper profile after the storm 1: Gravels on the storm berm and the surface layer of sand on the bar feature are observed. (*Source* [2]. Reproduced with permission of Elsevier)

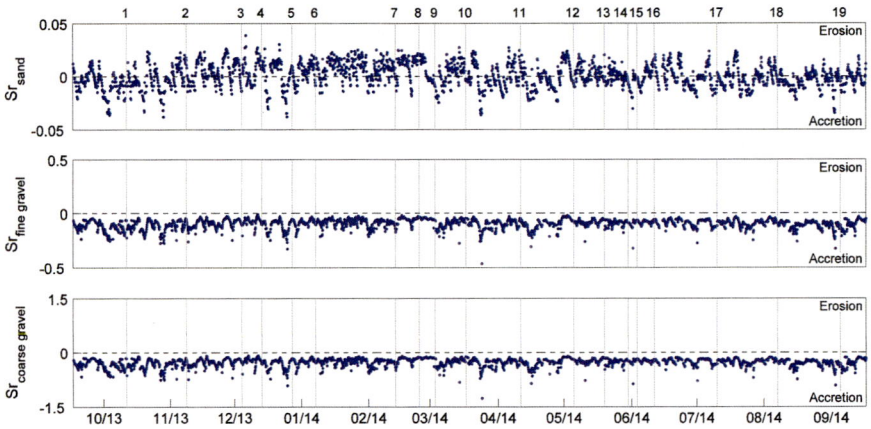

Fig. 5.9 Difference between the terms of erosion and accretion in the Eq. 5.1 [20] in study area 1. The three prevailing sediment sizes were considered: sand (*upper panel*), fine gravel (*middle panel*) and coarse gravel (*lower panel*). The vertical lines (*gray*) indicate the date of the field surveys. (*Source* [2]. Reproduced with permission of Elsevier)

gravel) in time and space (Fig. 5.10a), with average proportions of 31.7% sand and 68.3% gravel. The upper profile in study area 2 is similar to that of the study area 1, but the beach is wider (~50 m). The average sand-gravel ratio sampled across the entire upper profile for study area 1 is larger than for study area 2 (Table 5.3).

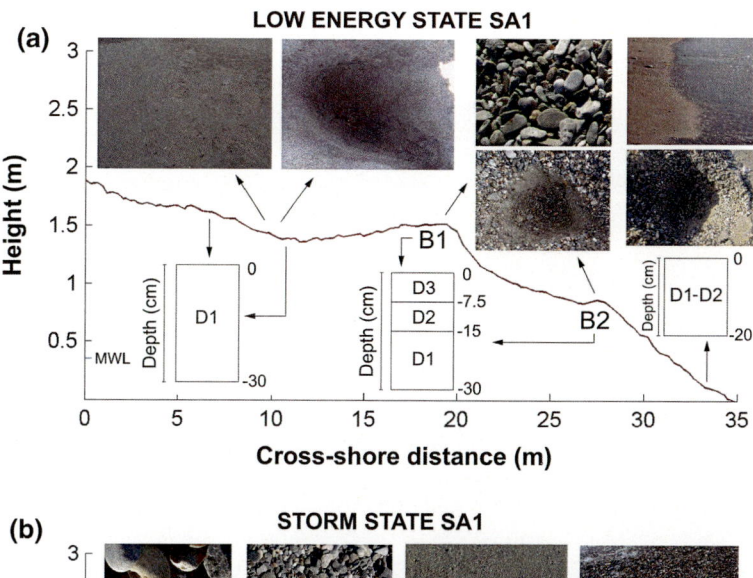

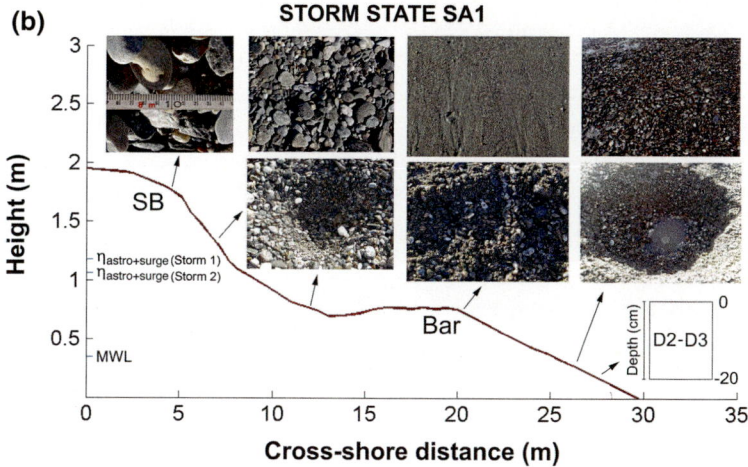

Fig. 5.10 **a** Low energy state in study area 1: morphology (including the berms B1 and B2), sedimentology and photographs. **b** State after storms in study area 1: morphology (including the storm berm -SB- and the bar feature), sedimentology and photographs. Height = 0 indicates the MLWS level. (*Source* [2]. Reproduced with permission of Elsevier)

5.5.2 Storm State

The two storms that occurred over the survey period (Fig. 5.10) induced marked changes to the beach profile. The low energy state berms were eroded and the upper beach profiles assumed a more concave shape (Fig. 5.10b). The storms also caused a decrease in the beach width of about 5 and 10 m in study areas 1 and 2, respectively. A storm berm developed on the upper part of the profile and a bar feature was generated in the lower part of the beach with a surface layer of sand over a gravel-dominated

Table 5.3 Sand-gravel percentages for the low energy and storm states on the backshore, foreshore and entire beach

		Low energy state (%)	Storm state (%)
Study area 1	Backshore	81.8 – 18.2	34.8 – 65.2
	Foreshore	31.7 – 68.3	24.8 – 75.2
	Entire beach	35.8 – 64.2	29.2 – 70.8
Study area 2	Backshore	80.8 – 19.2	31.1 – 68.9
	Foreshore	30 – 70	19.7 – 80.3
	Entire beach	33.1 – 66.9	23.7 – 76.3

substrate (Figs. 5.8b and 5.10b). In both study profiles, higher slopes were generally attained during storm conditions and the percentage of gravel increased by between 6 and 10 percent from the low energy state to the storm state (Table 5.3).

Considering the total number of samples taken after the two storms (18) and during the low energy states of the profile (27) in study area 1 and applying the *Student's t-test*, the result also confirm that the percentage of gravel is higher after storms (null hypothesis), with a significance level equal to 0.01. The same conclusion is drawn after applying the test in study area 2. These results are consistent with those obtained in Sect. 5.4: the finer material is selective transported offshore during storms, whereas under calm conditions the sand returns, covering most of the lag gravel (Fig. 5.10). This is a mechanism that differentiate MSGBs from sandy and pure gravel beaches [4, 12].

5.5.3 Transitional States

After the passing of storms, berms developed and progressively overlapped under the influence of low energy waves, contributing to the sediment variability both cross-shore and at depth (Fig. 5.11). The generation of berms represents a recovery trajectory, which is closely related to the total run-up (Figs. 5.3, 5.4 and 5.11), but this development of the berms can be interrupted at any one time by another storm. Figure 5.4 shows that the erosion/deposition rates were higher in the foreshore, where the measured sediment variability was also higher (Table 5.3). This is consistent with the conceptual model presented in Fig. 5.11, which suggests that the number of berms depends on the state of the profile and varies during the recovery process.

Figure 5.12a depicts the contribution of the wave run-up (vertical axis) and the sum of astronomical tide and storm surge (horizontal axis) to the unit volume variation (circles). Before the replenishment, it is observed that when the total run-up elevation was higher than the height of the upper berm ($\sim$1.52 m, Fig. 5.10a), the upper profile lost volume, whereas lower elevations (positive values of the free-board) increased the volume of the beach (Fig. 5.12a). Hence, beach erosion took place not only during

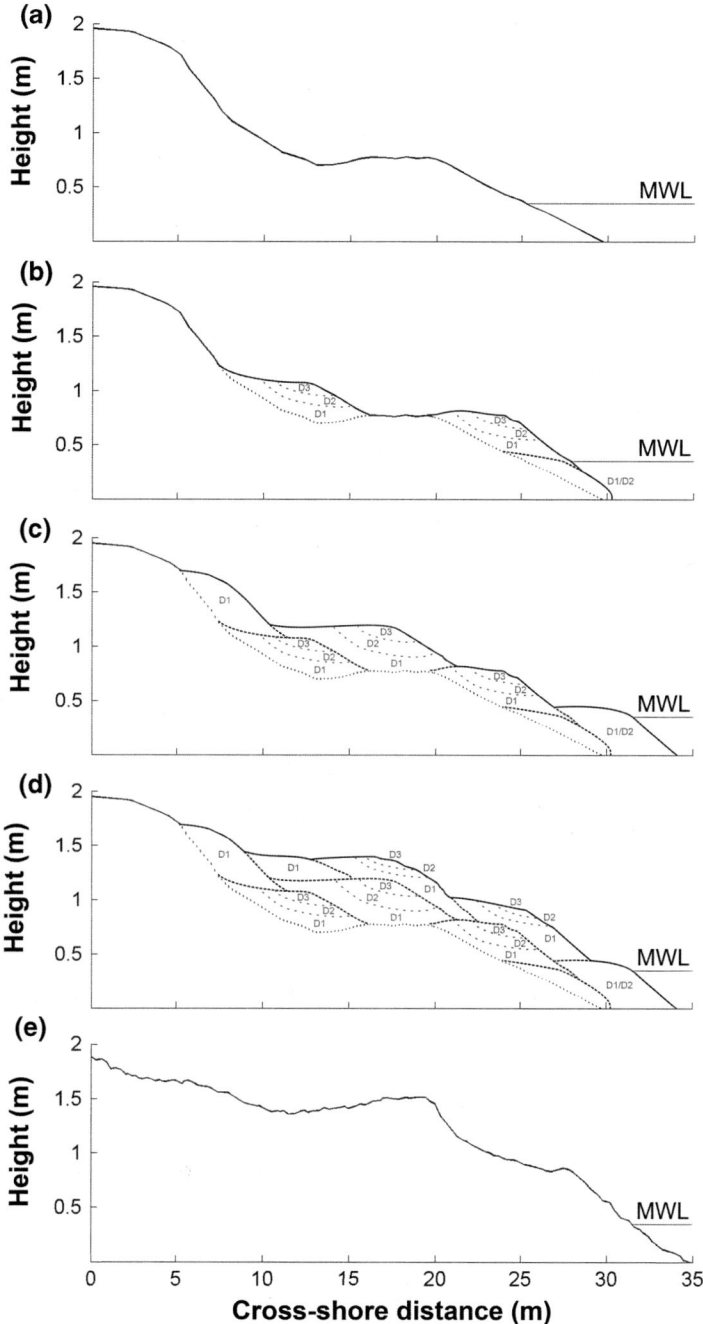

Fig. 5.11 Conceptual model describing the beach's accretionary response in study area 1. From *top* to *bottom*: storm (S), transitional-storm (T-S), transitional (T), transitional-low energy (T-L) and low energy (L) states. The number of berms depends on the state of the beach profile. (*Source* [2]. Reproduced with permission of Elsevier)

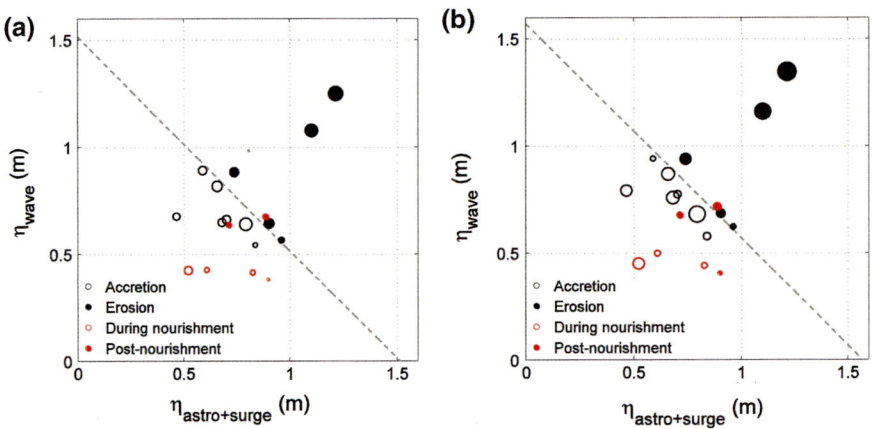

Fig. 5.12 Relationship between the wave run-up and the water-level elevation in study areas 1 **a** and 2 **b**. *Black circles* indicate accretion (unfilled) and erosion (filled) before the replenishment, whereas red circles indicate the artificial accretion (unfilled) and erosion (filled) afterwards. The size of the circles is proportional to the sediment volume change and the dashed gray line represents a total run-up of 1.52 m (**a**) and 1.58 (**b**). (*Source* [2]. Reproduced with permission of Elsevier)

both storms, when overwashing of the entire beach occurred; but also before surveys 2, 6 and 11, when the upper berm was overwashed.

Although the total run-up was similar in both study areas and the erosion/accretion behaviour of the upper profile was equal (Fig. 5.4, middle panel), the threshold elevation in study area 2 was ~1.58 m (Fig. 5.12b), coinciding with the height of the upper berm at this profile. These results indicate that the overwash process plays a key role in the beach dynamics, which is in agreement with previous works, such as [14]. Hence, other variables apart from wave height, such as pressure gradient and wind velocity, are essential in the evolution of the profile on micro-tidal beaches and the total run-up is demonstrated to be a more accurate threshold than wave height to differentiate between erosional and depositional conditions.

The evolution of both study areas was similar during the study period (Figs. 5.4 and 5.12), indicating that they are representative of the beach behaviour of Playa Granada. Differences between erosion/deposition rates and changes between low energy and storm states were higher in study area 2, probably because it experiences more energetic wave conditions (Fig. 5.2, second panel). The recovery rates at the two study areas seem to be related to the incoming wave directions: when the beach was subjected to westerly waves (surveys 2–3 and 11–12, Fig. 5.2), the accretion rates of study area 2 were higher than those of study area 1, whereas under prevailing easterly ones (surveys 9–10 and 12–13) the latter recovered faster. This could be due to the different longshore sediment transport gradients depending on the incoming wave direction, detailed by [1].

5.6 Conclusions

1. The generation and subsequent overlapping of berms is responsible for the sediment variability cross-shore and at depth on MSGBs. The cross-shore locations of these berms are related to the total run-up, as berms are modified by swash action. Thus, waves play a main role in the beach recovery.
2. The recovery of MSGBs has been shown to occur at a faster rate than on sandy beaches. This quicker recovery on MSGBs acquires importance for the design of strategies to help mitigate or adapt to global coastal erosion problems and the action of sea level rise.
3. The formation of a storm berm, the more concave shape of the upper profile and the increase in the percentage of gravels after storms all indicate reflective behaviour of MSGBs during high-energy conditions, and is dominated by the gravel fractions due to the selective removal of the finer material.
4. Total run-up elevations that exceed the height of the upper berm generate erosion, whereas lower elevations increase the unit volume of the upper profile representative of beach accretion. Hence, the total run-up represents a more accurate threshold dictating beach response than wave height.
5. The upper profile was flattened following an artificial input of sediment over June 2014 with different grain size distribution and lower D_{50} than the natural sediment. Three months after the nourishment, and in the absence of significant storms, the upper profile lost between 6 and $10\,m^2$. The beach volumes in September 2014 were similar to those measured in October 2013, showing the intervention was not effective.

Acknowledgements This work was partially supported by the Spanish Ministry of Economy and Competitiveness (Projects CTM2012-32439 and BIA2015-65598-P). The second author was funded by the Spanish Ministry of Economy and Competitiveness (Research Contract BES-2013-062617 and Mobility Grant EEBB-I-15-10002). The authors are indebted to Gerd Masselink for his valuable suggestions and comments. We also thank Servicio Provincial de Costas (Granada, Spain) for providing information about the artificial replenishment of the beach and Miguel A. Reyes-Merlo for his support with the field surveys.

References

1. Bergillos, R.J., López-Ruiz, A., Ortega-Sánchez, M., Masselink, G., Losada, M.A.: Implications of delta retreat on wave propagation and longshore sediment transport—Guadalfeo case study (southern Spain). Mar. Geol. **382**, 1–16 (2016)
2. Bergillos, R.J., Ortega-Sánchez, M., Masselink, G., Losada, M.A.: Morpho-sedimentary dynamics of a micro-tidal mixed sand and gravel beach, Playa Granada, southern Spain. Mar. Geol. **379**, 28–38 (2016)
3. Bowden, K.F.: Physical Oceanography of Coastal Waters. Ellis Horwood Ltd, Chichester, England (1983)

4. Bramato, S., Ortega-Sánchez, M., Mans, C., Losada, M.A.: Natural recovery of a mixed sand and gravel beach after a sequence of a short duration storm and moderate sea states. J. Coast. Res. **28**(1), 89–101 (2012)
5. Dean, R.G., Dalrymple, R.A.: Coastal Processes with Engineering Applications. Cambridge University Press (2002)
6. Folk, R.L.: Petrology of Sedimentary Rocks. Hemphill Publishing Company, Austin, Texas (1980)
7. Holthuijsen, L.H., Booij, N., Ris, R.C.: A spectral wave model for the coastal zone. In: Ocean Wave Measurement and Analysis, pp. 630-641. ASCE (1993)
8. Ivamy, M.C., Kench, P.S.: Hydrodynamics and morphological adjustment of a mixed sand and gravel beach, Torere, Bay of Plenty, New Zealand. Mar. Geol. **228**(1), 137–152 (2006)
9. Lee, G.h., Nicholls, R.J., Birkemeier, W.A.: Storm-driven variability of the beach-nearshore profile at Duck, North Carolina, USA, 1981–1991. Mar. Geol. **148**(3), 163–177 (1998)
10. Lesser, G.R.: An approach to medium-term coastal morphological modelling. UNESCO-IHE, Institute for Water Education (2009)
11. Lesser, G.R., Roelvink, J.A., Jatm, V.K., Stelling, G.S.: Development and validation of a three-dimensional morphological model. Coast. Eng. **51**(8), 883–915 (2004)
12. Mason, T., Coates, T.T.: Sediment transport processes on mixed beaches: a review for shoreline management. J. Coast. Res. **17**(3), 645–657 (2001)
13. Mason, T., Voulgaris, G., Simmonds, D.J., Collins, M.B.: Hydrodynamics and sediment transport on composite (Mixed Sand/Shingle) and sand beaches: a comparison. In: Coastal Dynamics, pp. 48–57. ASCE (1997)
14. Matias, A., Blenkinsopp, C.E., Masselink, G.: Detailed investigation of overwash on a gravel barrier. Mar. Geol. **350**, 27–38 (2014)
15. Matias, A., Masselink, G., Castelle, B., Blenkinsopp, C.E., Kroon, A.: Measurements of morphodynamic and hydrodynamic overwash processes in a large-scale wave flume. Coast. Eng. **113**, 33–46 (2016)
16. Matias, A., Masselink, G., Kroon, A., Blenkinsopp, C.E., Turner, I.L.: Overwash experiment on a sandy barrier. J. Coast. Res. **65**, 778–783 (2013)
17. Nielsen, P., Hanslow, D.J.: Wave runup distributions on natural beaches. J. Coast. Res. **7**(4), 1139–1152 (1991)
18. Ortega-Sánchez, M., Lobo, F.J., López-Ruiz, A., Losada, M.A., Fernández-Salas, L.M.: The influence of shelf-indenting canyons and infralittoral prograding wedges on coastal morphology: the Carchuna system in Southern Spain. Mar. Geol. **347**, 107–122 (2014)
19. Scott, T., Masselink, G., O'Hare, T.A., Davidson, M., Russell, P.: Multi-annual sand and gravel beach response to storms in the southwest of England. In: Proceedings of the 8th Coastal Sediments (2015)
20. Sunamura, T., Takeda, I.: Landward migration of inner bars. Mar. Geol. **60**(1), 63–78 (1984)
21. Thom, B., Hall, W.: Behaviour of beach profiles during accretion and erosion dominated periods. Earth Surf. Proc. Land. **16**(2), 113–127 (1991)
22. Wentworth, C.K.: A scale of grade and class terms for clastic sediments. J. Geol. **30**(5), 377–392 (1922)